PASS YOUR AMATEUR RADIO TECHNICIAN CLASS TEST – THE EASY WAY
2014-2018 Edition

By: Craig E. "Buck," K4IA

ABOUT THE AUTHOR: "Buck," as he is known on the air, was first licensed in the mid-sixties as a young teenager. Today, he holds an Amateur Extra Class Radio License.

Buck is an active instructor and a Volunteer Examiner. The Rappahannock Valley Amateur Radio Club named him Elmer (Trainer) of the Year three times and Buck has successfully led many students through this material.

Email: K4IA@arrl.net

Published by EasyWayHamBooks.com
130 Caroline St. Fredericksburg, Virginia 22401

This and other "Easy Way" Books by Craig Buck are available at Ham Radio Outlet stores and Amazon:
"Pass Your Amateur Radio General Class Test– The Easy Way"
"Pass Your Amateur Radio Extra Class Test– The Easy Way"
"How to Chase, Work & Confirm DX– The Easy Way"
"How to Get on HF – The Easy Way"

ISBN 978-1508738220

Library of Congress Control Number PENDING .8

PASS YOUR AMATEUR RADIO TECHNICIAN CLASS TEST – THE EASY WAY

TABLE OF CONTENTS

INTRODUCTION 4

THE TEST 4

HOW TO STUDY...9

HOW I GOT STARTED 10

INTRODUCTION TO AMATEUR RADIO 19

WHO ARE YOU? 22

WHERE CAN YOU OPERATE? 25

WHAT DO WE TALK ABOUT? 26

WHO CAN YOU TALK TO? 28

WHO IS IN CONTROL? 29

BANDS AND FREQUENCIES................................... 31

MODES 36

    FM MODE...36

    AM MODE 37

    SINGLE SIDEBAND MODE (SSB)...................... 38

    CW MODE (MORSE CODE) 39

BANDPLANS 41

PROPAGATION.. 42

BE SAFE, STAY SAFE 46

RADIO OPERATION 50

HOW TO START A CONVERSATION 53

EXTENDING YOUR RANGE WITH REPEATERS.......... 54

OPERATING SIMPLEX... 58

EMERGENCY.. 59

SATELLITES 62

COMPUTERS..65
RADIO DESIGN..67
CLEAN UP THE SIGNAL 69
ELECTRONIC THEORY 73
VOLTS, OHMS, AMPERES – OHM'S LAW 76
    POWER IN WATTS 78
HOW TO DRAW A RADIO 79
DECIBELS...83
MOVING DECIMALS 84
BUILDING EQUIPMENT & MEASURING VALUES........85
ANTENNAS..86
    POLARIZATION 86
    ANTENNA LENGTHS...................................87
    DIRECTIONAL ANTENNAS 88
    COAXIAL CABLES 89
    ANTENNA ANALYZERS AND SWR 91
BATTERIES 94

QUICK SUMMARY

INTRODUCTION TO AMATEUR RADIO 96
INTERNATIONAL REGULATIONS 97
WHO ARE YOU? 98
WHO IS IN CONTROL? 99
IS THAT ALLOWED?.. 100
BANDS AND FREQUENCIES............................ 102
TECH PRIVILEGES 103
BE SAFE, STAY SAFE 104
RADIO OPERATION....................................... 107
EMERGENCY.. 110
PROPAGATION... 111
MODES.. 115
    FM FREQUENCY MODULATION 115
    SSB SINGLE SIDEBAND MODULATION.............. 115

CW MORSE CODE.. 116
RADIO DESIGN.. 117
CLEAN UP THE SIGNAL 118
ELECTRONIC THEORY 120
    VOLTS.. 120
    AMPERES.. 120
    RESISTANCE 120
    INDUCTORS 121
    CAPACITORS... 121
    TRANSISTORS 122
    DIODES.. 123
VOLTS, OHMS, AMPERES – OHM'S LAW 123
    POWER IN WATTS 125
HOW TO DRAW A RADIO 125
DECIBELS.. 128
MOVING DECIMALS 129
BUILDING EQUIPMENT & MEASURING VALUES 130
ANTENNAS... 130
    POLARIZATION 130
    DIRECTIONAL ANTENNAS.............................. 131
COAXIAL CABLES... 132
ANTENNA ANALYZERS AND SWR 133
BATTERIES 134

INDEX.. 136

INTRODUCTION

There are many books to help you study for the Amateur Radio exams. Most focus on taking you through all the questions and possible answers on the multiple-choice test. The problem with that approach is that you must read three wrong answers for every one right answer. That's 1,278 wrong answers and 426 right answers. No wonder people get overwhelmed.

This book is different. There are no wrong answers. I'm going to take you on a journey of discovery. In the process of relating my personal journey through Amateur Radio, I'm going to answer every question on the Amateur Radio Technician Class exam in a way that will help you understand the answers and be entertained at the same time. **The test questions and answers are in bold print to help you focus.**

I might repeat a question because it fits into the narrative in more than one place. That will just reinforce the answer in your mind. Please excuse the tortured grammar and syntax of the bold print answers. I am copying them word-for-word from the exam.

The second part of this book is a Quick Summary – just the question and the right answer. I've condensed what others take over 200 pages to say into 40 pages because you do not see the wrong answers.

Once you get your Technician license, you'll want to move up with my other books: "How to Pass Your Amateur Radio General Class Test – The Easy Way" and "How to Get on HF – The Easy Way." Order them today so you'll be ready! EasyWayHamBooks.com

THE TEST

The test itself is 35 questions out of a possible pool of 426. The pool is large but only about one in a dozen questions in the pool will show up on your exam, one from each of the test subjects. I can guarantee you won't get 35 questions on Ohm's law. Many questions ask for the same information in a slightly different format, so there aren't 426 unique questions.

There are **three license classes: Technician, General, and Extra** each conferring more privileges. So, how hard is the Technician test? It is not hard. There is no heavy-duty algebra, calculus or trigonometry required for this test. Don't let the material intimidate you. There is no Morse code required for any class of Amateur license. Morse code is still very much alive and well on the amateur bands, and there are many reasons you might want to learn it in the future. Just don't worry about it for now.

Billy and I studied for our test by reading textbooks. We didn't know what questions would be on the test. Now, the questions and answers are all available so you can study with confidence that you are focusing on the right material.

The questions and the answers on your test are word-for-word out of the published question pool. The order is scrambled so answer "a" in the question pool is not necessarily "a" on your test. Don't try memorizing the multiple-choice letters.

I have personally taken students who didn't know anything about electronics, radio or math through this material, and they passed. You want to learn it all but if there's a question you just can't get, skip it. Chances are it won't be on your test. And, you can miss nine out of the thirty-five questions and still pass.

THE TEST

Do you know what they call the person who graduates last in his class from medical school? "Doctor" – the same as the guy who graduated first in the class. My point is: all you need is to pass, and no one will know the difference.

The questions are multiple-choice, so you don't have to know the answer, you just have to recognize it. That is a tremendous advantage for the test-taker. However, it's difficult to study the multiple-choice format because you can get bogged down and confused by seeing three wrong answers for every correct one. I think that makes it harder to recognize the right answer when you see it.

The best way to study for a multiple-choice exam is to concentrate on the correct answers. The second part of this book is a condensed version of the test with only the correct answers. When you take the test, the correct answer should jump right out at you.

If you don't know the answer, try to eliminate the obviously wrong answers. Then guess. There is no penalty for guessing wrong, and you greatly increase your odds by eliminating the bogus responses.

Long ago, when my friend, Billy, and I sat for our General test, we took a bus to the FCC Building in Washington, DC. There we met the sternest proctor I have ever suffered under. Picture the short sleeve white shirt, skinny black tie, pocket protector, slide-rule-on-the-belt, thick eyeglasses, cigarette smoke and a full ashtray. I felt that if I did not pass the test, I was going straight to jail. I was sure the FBI told him I had listened to Radio Moscow.

Today, it is all much easier. Volunteer Examiners organized under the authority of the FCC, administer Amateur Radio testing. VEs are local hams who devote their time to help you get your license. In most areas you can find a VE group testing at least once a month. Check on the licensing tab of ARRL.org to find a test site. The modest exam fee reimburses the VE's costs. No one profits except you.

On test day, you should bring a picture ID, your Social Security Number (not the card, just the number)[1] a pen and pencil and about $15. You need to ask ahead to get the exact amount and see if they prefer cash or check. If you bring a calculator, you must clear the memories. Turn off your cell phone and don't look at it during the exam.

You will fill out a simple application (FCC Form 605), and you might save a few minutes by downloading the form from the FCC website and filling it out ahead of time. Your Volunteer Examiner (VE) team will select a test booklet for you. Make sure you ask them for the easy one – that's good for a cheap laugh. They have no idea which questions are in which booklet. You also get a multiple-choice, fill-in-the-circle answer sheet. The answer sheet has room for 50 questions, but you stop at 35. You can take notes and calculate on the back of the answer sheet – not in the booklet. VEs re-use the booklets.

The VE team will grade your exam while you wait and give you a pass/fail result. Don't ask them to go over the questions or tell you what you missed. They don't know and don't have time to look. When you pass, there is a bit more paperwork, and you walk out with a CSCE – a Certificate of Successful Completion of

---

[1] Many VEs request you obtain a Federal Registration Number (FRN) from the FCC instead of using your Social Security Number. The FRN is used only for your communications with the FCC. Search "FRN FCC" to get the link to the FCC website where you can register. Then, bring your FRN number to the test session.

THE TEST

Examination. The team processes your Form 605 and CSCE and your call sign appears in the FCC (Federal Communications Commission) database in a few days. Congratulations.

You can start transmitting when you see your call sign appear online in the FCC database. This takes about a week.

Your license is good for 10 years and can be renewed without another test.

Also, there is a two-year grace period so you can renew your license without another test.
You may not transmit during the grace period because you are not licensed.

Once you get your license, you want to make sure to keep the FCC updated on any address changes. **When correspondence from the FCC is returned as undeliverable because the grantee failed to provide the correct mailing address, revocation of the station license or suspension of the operator license may be the result.**

And while we're on the subject of the FCC, you should know that **the FCC has the power to inspect your station and records at any time.** It is highly unlikely that you would ever get a visit from the FCC, but if it happens, you need to cooperate.

HOW TO STUDY

You ace a multiple-choice exam by learning to recognize the right answer and eliminate wrong answers. You could study by poring over the multiple-choice questions. That has been the traditional approach taken by most classes and license manuals. The problem with that method is you have to read through three wrong answers to every question. That is both frustrating and confusing. Why study the WRONG answers?

The approach here is that you never see the wrong answers so the right answer should pop out at you when you see it on the test. You don't need to memorize the whole answer – just enough of it to recognize. Even if you don't understand a word of the question or answer, you will recognize the right answer.

I don't recommend you take practice exams until the week before the test after you feel you have mastered the material. The reason I recommend waiting to take the practice exams is that I don't want you to get confused by seeing all the wrong answers. I want you to recognize the right answers first.

Your Amateur Radio license is called a "ticket", and that is just what it is. Getting a ticket is the beginning of the journey. You can't get on an airplane and start your journey without a ticket. You can't get on the air and start your journey without a ticket. The ticket, by itself, is just a piece of paper. However, it is your travel pass to new places. Your Ham ticket can take you places you never thought you would go. Get your ticket first. The rest will come later.

Throughout the material, the correct answer is in **bold print**. Hints, notes, and cheats are in *italic*.

HOW I GOT STARTED

Begin at the beginning. I grew up in the days before personal computers, the internet, video games and 500 channels of color TV. We made our entertainment. I remember building an airplane cockpit using shoe boxes with toothpicks for control levers. We shot marbles and built elaborate obstacle courses that would make a golf course designer proud. If ants got in the way, we dumped lighter fluid down their hole and lit it. No mercy. That was just third grade.

These were the days when you rode a bike without a helmet; cars didn't have seatbelts, and you stayed out until the street lights came on. We played in the woods, caught snakes and lit cherry bombs. Model boats and planes filled my shelves. In fifth-grade science class, we built a crystal radio set. It only tuned one station. At night, I would tuck the earpiece under my pillow and fall asleep listening to basketball games. Basketball on the radio took a lot of imagination and held no interest for me. I was just amazed to hear radio coming from a rock and no batteries. That station was about 30 miles away.

Tuning the family's Crosley 5-tube AM radio, I discovered "skip" – the bouncing of radio waves off the ionosphere. **Radio waves are electromagnetic. The ionosphere is the part of the atmosphere that enables propagation.** Radio waves bounce off the ionosphere and come down a distance away. It is possible to hear stations hundreds or thousands of miles off as you have probably discovered at night with your car's AM radio.

Around 7th grade, I got a shortwave radio kit for Christmas. Building the kit, I learned about **schematic symbols which stand for the various components.** I burned my fingers more than once learning to solder. **We use rosin core solder** because Dad's acid core plumbing solder would corrode the joint. I also learned that cold solder joints are no good. **Cold solder joints look grainy and dull** and make a bad connection. You need to reheat the joint and try again. They are caused by insufficient heat or moving the part before the solder has set.

Miraculously, the radio worked. There was something eerie about hearing Radio Havana Cuba or Radio Moscow during the cold war. I kept expecting the FBI to break down the door. I got my first lesson in political "spin" hearing Cold War propaganda even before the term was invented. I remember thinking it was funny that the tune used to identify Radio Havana was the same one the Gillette Razor Company used in their commercials. I guess since Castro had a beard, he didn't know about Gillette razors.

My buds and I raided junk parts from behind the TV repair shop. We'd strip the chassis to use for our own projects. We'll talk about those parts and what they do later.

Radio wasn't our only mischief. There were home-made black powder and the "arc." The phone company used 45-volt batteries as backup power and threw them away long before they died. Each one was composed of thirty individual AA size 1.5-volt batteries, hooked in series (end to end) and encased in a tar brick. We would grab them out of the trash bin for our nefarious purposes. These were **non-rechargeable carbon zinc batteries.**

Types of rechargeable batteries are:
Nickel-metal hydride
Lithium-ion
Lead-acid gel-cell
All the choices are correct.
Hint: Lots of rechargeable battery types so all are correct.

You can recharge a lead-acid battery by connecting it in parallel (across) a car battery and running the engine.

Lead-acid batteries are very dangerous because shorting the terminals can cause burns, fire or an explosion. Car batteries can deliver a tremendous amount of current in a short time – enough to melt a tool or the ring on your finger. **Lead-acid batteries can give off explosive gas if not properly vented. If you charge or discharge lead-acid batteries too quickly, they can overheat and give off flammable gas or explode.** The flammable gas is hydrogen. Remember the Hindenburg? An exploding car battery is also going to spray sulfuric acid.

Voltage is electromotive force. The basic unit is the volt. We could measure it with a voltmeter hooked up across (in parallel with) the circuit – from one end of the battery to the other.

You do have to be careful when measuring high voltages that the meter and leads are rated at the voltages to be measured.

We knew that two of the 45-volt phone company batteries in series (end to end) are 90 volts, wouldn't it be fun to hook twenty 45 volt batteries in series for 900 volts? Why stop there? We have lots of batteries. I am not sure where we stopped, but we had too many batteries for our own good.

**Current flowing through a body will cause a
health hazard by:
Heating tissue
Disrupting electrical functions
Causing involuntary muscle contractions
All of the choices are correct**

My buddy, Billy, struck the two wires together and
pulled them apart to draw a flaming, spewing, spitting
and smoking foot-long electrical arc. **The flow of
electrons is called current and it is measured in
amperes.**

We could have **measured the amperes (amps)
with an ammeter connected in series (in line)
with the circuit.** You measure flow through the
circuit.

Copper is a good conductor, but there is some
resistance to the flow of current even in copper wire.
A good insulator like glass has a very high
resistance and won't pass current.

**Resistance is measured in ohms with an
ohmmeter.**

The resistance of the wire caused the wires to
consume some of the power, and the wires got so hot
Billy had to drop them. The batteries heated up as
well, and the entire crate of batteries melted in a pool
of hot tar. We should have used a fused line. **A Fuse
or Circuit Breaker in the line interrupts the
power in case of overload. Don't put a 20 amp
fuse in place of a 5 amp fuse or excessive
current could cause a fire.**

The pool of hot tar was nothing compared to the spark
that flew into a large jar of our homemade black
powder. I could see my short life passing before my
eyes. All the ingredients to make gunpowder were in

HOW I GOT STARTED

our local hardware store, but we had never really perfected the recipe to make it explosive. I remember the hardware-store clerk glaring at us with that "I know what you boys are up to" look. I think he knew I had listened to Radio Moscow.

It was fortunate the black powder did not explode, but now we had a Fourth-of-July fountain erupting in Billy's basement. The story has a happy ending: we didn't electrocute ourselves, the house did not burn down and, to my knowledge, Billy's parents never found out. Although, now that I think about it, I wasn't invited over for a while. That was probably seventh or eighth grade.

We were experimenters and builders – kids with curiosity. Today, we would be called "makers." What about ham radio? What was the lure? Simple. It was the first time all the stuff they were trying to teach me in school had a use. How many times have you heard adolescents complain, "I don't need to know algebra." "I'll never use chemistry." "Why do I have to learn about the ionosphere?" "Who cares where Pitcairn Island is?" And, on and on.

If you want to know how long to make an antenna, you use a little algebra. If you want to know the best time to talk to Australia, you need to understand the ionosphere. Battery chemistry will help you decide the best power source for your equipment. If you know geography, you know which way to turn your directional antenna. If you know languages, you can say "hi" to a foreigner in his language. If you know astronomy, you can appreciate meteor scatter and sun spots or know when to bounce a signal off the moon or satellite. If you learn physics, you can design an antenna. If you know history, you can appreciate Pitcairn Island. I could teach most of an entire high school curriculum based on Amateur Radio alone.

And what does Pitcairn Island have to do with it? Plenty. During the summer, Mom would take me to the library. I enjoyed the tall-ship swash-buckling tales with pirates and mutinies. My favorite was The Mutiny on the Bounty – a true story from 1789 with real people and a real mutiny with real villains and heroes. It is up to you to decide who the villains were and who the heroes were.

The mutineers, led by Fletcher Christian, fled to Pitcairn Island in a remote part of the Pacific Ocean. One night, tuning the radio, I heard VR6TC, Tom Christian, a direct descendant of Fletcher. VR6TC was Tom's call sign at the time. I "worked" him. That's ham-speak for "we made contact." I was shaking.

Tom was famous because of the Bounty story and the rarity of his location. I never heard Tom on the air again, and he became a Silent Key[2] a few years ago. Pitcairn was so remote and isolated; there were no other means to communicate with the outside world. That is why Tom and his wife had their licenses.

Monk Apollo in the Holy Monastery of Docheiariou at Mt. Athos (Greece) has his license for the same reason. The monastery was founded in the 10th century and is closed to the world. When one of the monks fell ill, there was no way to summon help. Monk Apollo obtained an Amateur Radio license to provide emergency communication. Talking to Monk Apollo is another great thrill. See what you can learn from ham radio?

I wrote a "What I did last summer" essay about contacting Tom Christian for English class. My teacher

---

[2] A deceased ham is referred to as a Silent Key, a throwback to the days of telegraph when the deceased's Morse Code key went silent. You will see the call sign listed VR6TC (SK).

HOW I GOT STARTED

didn't believe me. She said the essay wasn't
supposed to be fiction.

Months later, I received a QSL[3] card from Tom. I
enjoyed rubbing her nose in it. Maybe that's why she
gave me a C in English that quarter. Gloating is never
a wise move.

Somehow, over the years, Tom's QSL card, VR6TC,
was lost, but you can see it on the wall to the top left
of the map in my picture. The picture is from 1967.

The black box in the picture of my station is the
receiver and the gray box to the right is the
transmitter I built from a kit. Today, most hams use
**transceivers: a transmitter and receiver in one
box.**

---

[3] Hams use Q signals as abbreviations. A QSL card is a postcard hams send
each other as a written confirmation of their contact.

Here is Monk Apollo's card from Mount Athos. It is one of the rarer ones in my collection.

K4IA, VR6TC, SY2A – what does it all mean? The country's communications agency issues call signs. In our case the **FCC, Federal Communications Commission regulates and enforces the rules. Worldwide, regulation is through the International Telecommunications Union (ITU) which is a United Nations agency for information and communication technology issues**. Each country is assigned prefixes so you can tell a person's location from the prefix. **The test asks you for a valid US call sign, and the answer is W3ABC**.

You use the call sign to identify yourself, in English, every 10 minutes and at the end of your conversation.

The only exception is if you are using Amateur Radio to control a model craft. Transmitting identification would confuse the model craft. It is expecting to hear only chirps and beeps. **To identify, you put a label on the craft with your name, call sign and address. You are limited to 1 watt of power,** but that will take your model a long

HOW I GOT STARTED

way. Don't let your drone go over the fence at the White House.

My favorite mode was, and still is, CW, Morse Code. **The code we use is called International Morse Code.** I liked it because no one could tell I was a kid. My squeaky voice was a dead giveaway on voice modes.

My call sign at the time was WA4TUF. When I was re-licensed in 1999, the FCC gave me KG4CVN – a mouthful and laborious to send on CW. C and V sound alike,so I always **used the phonetic alphabet to identify**.

Fortunately, the FCC has a vanity license program, and **any licensed amateur can apply for a different call sign.** Certain styles are reserved for different classes. **K1XXX is an acceptable call sign for the Technician Class**. One letter, one number, three letters. Remember, **K1XXX.**

INTRODUCTION TO AMATEUR RADIO

We're all familiar enough with radio. Radio is so common today that we take it for granted. We have AM/FM and satellite radios in our car. We use cell phones and GPS navigation systems. We've come to depend on wireless Internet, wireless doorbells and wireless computer accessories. But, when Guglielmo Marconi transmitted a signal across the Atlantic Ocean in 1901, it was a remarkable feat that made world headlines. Marconi was essentially an experimenter and Amateur Radio today is a continuation of that experimenter spirit.

When Titanic sank in 1912, the world awoke to the value of radio. Stations in Europe and the United States heard Titanic's powerful distress calls, but the radio operator on the closest ship had gone to bed and turned off his equipment. There was no standard procedure for handling distress calls at the time.

The answer on today's tests is that the purpose of Amateur Radio is **advancing skills in the technical and communication phases of the radio art.** *Hint: Radio art? That odd phrase should stick with you as the right answer.* **Another purpose is to enhance international goodwill. A permissible use is to conduct radio experiments and communicate with others around the world.**

Nobody knows where the term 'ham radio" came from, but you'll often hear it used to describe the Amateur Radio service. I suspect people use "ham" because they have a hard time spelling amateur.

An amateur station is a station in Amateur Radio Service consisting of the apparatus necessary for

carrying on radio communications. That is a circular definition you should recognize.
Hint: Recognize – don't memorize!

It is from PART 97 of the FCC rules.   PART 97 is the part of the FCC rules governing Amateur Radio.

Governments have long recognized the value of a cadre of experienced radio operators. There are almost 700,000 amateurs in the United States alone. That is a tremendous number of knowledgeable radio operators ready to bring their equipment and expertise to public service or emergency uses. All this expertise and all this equipment didn't cost the U.S. government a dime.

You can appreciate the need to regulate the radio spectrum. Without some form of regulation, your radio would sound like a raucous cocktail party and communication would be very difficult.
Worldwide, regulation is through the International Telecommunications Union (ITU) which is a United Nations agency for information and communication technology issues.
Hint: Don't memorize this – recognize ITU/United Nations

A US station can operate in international waters on a US flagged vessel.

The are three ITU regions in the world with slightly different rules in each region. You follow the rules of your location. **You might be in a different ITU region and be subject to those rules.** The United States happens to be in region 2. **Some US territories are outside Region 2 and can have different frequency assignments.**

Some countries may notify the ITU they object to communication altogether and you are not allowed to talk to them even if you hear one (which you won't). North Korea is an example. If you hear someone using a call that starts with P5, it is probably a bootlegger – someone who is making up a North Korean call sign.

Amateurs are licensed to operate on frequencies that can easily travel around the world. Citizens band is limited to 5 watts of power (about the same as a night light). Amateurs can operate with up to 1,500 watts. That is a big responsibility, and you need to prove you are capable of handling it. There are three classes of Amateur Radio licenses. The entry level is called the Technician Class.

The Technician Class license authorizes you to transmit mainly on the: **VHF (Very High Frequency 30 – 300 MHz)** and **UHF (Ultra High Frequency 300 – 3000MHz)** bands which are suitable for local communication.

(HF, High Frequency: 3 – 30 MHz) bands are sometimes called "short-wave" and are suitable for worldwide communication." Techs have some limited privileges on HF. We'll deal with frequencies and bands later but let's keep it simple for now. The dividing lines are **3-30, 30-300 and 300-3000 for High, Very High, and Ultra High.**

The General Class license authorizes you to transmit on all the bands. Extra Class licensees A get a small bit of additional spectrum within the bands. The Extra Class license gives full amateur frequency privileges.

WHO ARE YOU?

You will receive an Amateur Radio call sign from the FCC. What is that? You've heard commercial radio stations identify themselves using call signs. For instance, WMAL is a popular radio station in Washington, DC. Radio amateurs also receive a unique call sign issued by the FCC. The International Telecommunications Union (ITU) allocates call sign groups to different countries. US call signs all began with a letter W, K, N, or A signifying that we are in the United States

You only need to know that **W3ABC would be an example of a valid US call sign.** I am going to use that call throughout this book so you will remember. If you heard an amateur whose call sign began with DL, you would know he is in Germany. A call sign beginning with PY would be from Brazil. There are 340 recognized "countries" each with a unique call sign prefix. I am very proud that I have confirmed contacts with 333 of the 340.

The FCC generates your first call sign in sequential order. When Billy and I got our licenses, we took the test together and received consecutive call signs. He was WN4TUE, and I was WN4TUF.

The United States is divided into ten numbered districts. The "4" in my call sign signifies that it was issued for the United States fourth call district which is the southeast. The call sign **W3ABC** would be the third call district, mid-Atlantic. The district numbering rules have become somewhat blurred and only apply to the original issue. If you move from one area to another, you can still keep your call sign.

If you are operating out of your area, you can use a self-assigned indicator. Examples might be KL7CC/W3. I know he is from Alaska because his call begins with KL7. The W3 tells me he is operating from the mid-Atlantic area which is the third call district. **On voice, it would sound like "KL7CC stroke W3" or "KL7CC slant W3" or "KL7CC slash W3."** Stroke, slant, slash – it's all the same.

When you pass your General test, you can use the General bands even though your new class has not yet appeared online. **While waiting for your upgrade to appear in the FCC database, you are required to sign /AG after your call**. *Hint: Awaiting General.*

Sometimes you'll hear a call sign with only **one letter in the prefix and suffix such as W7C. These are known as special-event call signs.** They are issued by the FCC on a temporary basis to commemorate a special event such as National Clothesline Day (I kid you not). **A temporary "1 by 1" format (letter – number – letter) call sign can be assigned for operations in conjunction with an activity of special significance to the amateur community.**

When our radio club operates in support of a public service event, we will often use what are called **"tactical" call signs. A station might identify as "Race Headquarters."** In these situations, it is more useful for everyone to know <u>where</u> you are, rather than <u>who</u> you are.

When using tactical identifiers, your station must transmit the FCC-assigned call sign every 10 minutes during and at the end of a conversation. *Hint: This rule never changes.*

Individuals get a call sign, but it is also possible to form a club and the club receive a call sign. **Club calls are administered by a trustee. It takes four individuals to form a club.** *Cheat: remember four to form*.

WHO ARE YOU?

You're required to identify with your call sign in English every 10 minutes during and at the end of your contact. The rest of the time you can speak any language you want.

You're not allowed to use secret codes.
The exception is that digital codes can be used to control a model airplane or spacecraft.

Note that Morse code isn't secret so it is acceptable. In fact, you can identify in Morse Code even if you are using another mode for your conversation. **If you are transmitting phone (voice) signals, you can identify with phone or CW (Morse Code).** You will hear repeaters identify in Morse Code.

Hams often use phonetic alphabet because it's difficult to understand the difference between letters through static crashes and interference on the radio. The letters C, E, D, etc. sound alike, but it is very easy to tell the difference between Charlie, Echo, and Delta. **Use of the phonetic alphabet is encouraged by the FCC when identifying your station when using phone.**

There are several standard phonetic alphabets, and you will get used to them. You might hear cutesy phonetics on the air, but they don't help with understanding. Kilo 4 Tango Sierra is easy to understand. Kilo 4 Tequila Sunrise is funny but not clear or easily recognized.

WHERE CAN YOU OPERATE?

In addition to operating in the US and territories, **your US Amateur Radio license entitles you to operate from any vessel or craft located in international waters and documented or registered in the United States.** I've worked plenty of /MM stations (maritime mobile). Some cruise lines will let you operate onboard (check first). Most of the /MMs I have worked have been freighter crew members. I have also worked stations that are /AM, meaning they are "aeronautical mobile" and operating from an airplane. Don't try this on a commercial airliner.

You are also authorized to operate your amateur station in a foreign country when the foreign country authorizes it. The United States has reciprocal operating agreements with many foreign countries so you can take your Amateur Radio station and license with you when you travel.

I took a radio to Israel in 2015. I went through security at Reagan National, JFK and with the Israeli carrier, ELAL. These are the toughest security checks in the world. I had the radio with coils of antenna wire and coax in my carry-on, and no one questioned what it was. On the way back, I put the gear in checked luggage with no issues. If you travel with your radio, take a copy of your license and a magazine ad or instruction manual showing what the equipment is – just in case.

Israel has a reciprocal agreement with the US, so I was able to operate 4X/K4IA and did not need to get a license from the Israeli government. I was required to have a copy of my US license, but no one asked to see it.

WHAT DO WE TALK ABOUT?

That's a question I get asked a lot. Typically, our conversations are very much like those you would have with any person you're meeting for the first time. We talk about common interests such as our equipment, the weather, how long we've been licensed, and other mundane **topics incidental to amateur service or of a personal nature.** It wouldn't be polite or proper to ask a foreign station when they're going to get rid of their bum dictator.

Locally, you might join a roundtable with fellow commuters. Some hams meet regularly and carry on long-distance conversations for years. They may never meet face-to-face but become part of an extended family.

Amateur radio isn't for business. Here's an example: I'm on my way over to my friend Bob's house to help him with an antenna. I call him on the radio and ask if he would like me to bring over pizza. He says, sure – whatever kind you like. Another ham, who also happens to own a pizza parlor, hears me and calls. "Hey Buck, what kind of pizza do you want. I'll have it ready for you." That's a no-no. Same pizza, different results.

There is an **exception for the incidental sale of Amateur Radio equipment.** Even that is supposed to be informal, and you cannot make a business of selling Amateur Radio equipment on the air.

We're amateurs and can't charge for our services. **An exception exists for classroom instruction at an educational institution.** The control operator is paid to be a teacher, not a ham.

Transmissions that contain obscene or indecent words or language are prohibited, but there is no

WHAT DO WE TALK ABOUT?

list of banned words. You are expected to know better. The big difference between licensed Amateur Radio operators and Citizens Band operators is that the hams are usually polite and hardly ever profane. You won't be afraid to let your family listen in on the conversations.

Music is only allowed when transmitted to the International Space Station. Most US astronauts are ham radio operators, and they will often get "on the air" while in space. They wake up to music sent over the ham radio link.

Hams aren't supposed to engage in "broadcasting." **Broadcasting means transmissions intended for reception by the general public** as opposed to transmissions intended for a particular station. **There are exceptions for code practice, information bulletins, and emergency communications**.

Emergencies are defined as involving the immediate safety of human life or property.

You are not allowed to intentionally interfere which means **to seriously degrade, obstruct or repeatedly interrupt others.**

WHO CAN YOU TALK TO?

Hams would normally talk to other hams as the frequencies we operate on are restricted to the Amateur Radio service. However, there is an exception. **During an Armed Forces Day Communications Test, an FCC-licensed amateur station may exchange messages with a US military station.** There are several hams operating out of Guantanamo Bay, Cuba, but they are operating as amateur stations not military. You can identify a Gitmo station because the call is KG4 and two letters.

Most countries allow ham radio operation, but **FCC-licensed amateur stations are prohibited from exchanging communications with any country whose administration has notified the ITU that it objects to such communications.** North Korea does not allow Amateur Radio. *Hint: Ditch the overcomplicated question. If a country objects, you can't talk to a ham there.*

There are some special rules for what is called "third party traffic." Third party traffic is when someone who is not a ham is speaking through your station. You might have a friend in your shack[4], and you allow him to say hello to another station.

The FCC rules authorize the transmission of non-emergency third-party communications to any foreign station whose government permits such communications. *Key phrase: "whose government permits."* Third party traffic is allowed within the United States but prohibited by some foreign countries. You don't need to memorize any lists, just know they are out there if the occasion arises and look it up.

---

[4] The radio shack on a boat is the room where the radios are located. "Shack" is the term used for the place you operate your radio.

WHO IS IN CONTROL?

There is a difference between a control operator and a control point. The control operator is a person, and the control point is a place.

An amateur station must have a control operator when the station is transmitting. Only a person for whom an Amateur Radio/primary station license grant appears in the FCC database is eligible to be the control operator of an amateur station. In other words, only a licensed ham can be a control operator.

The class of operator license held by the control operator determines the transmitting privileges of an amateur station.

Therefore, a Technician Class control operator may only operate with those privileges allowed to Technician Class licensees. However, if an Extra Class licensee was acting as the control operator, the Technician Class licensee could operate a radio anywhere allowed the Extra Class license. That is because the Extra Class licensee is "in control" of the station.

A Technician Class licensee may never be the control operator of the station operating an exclusive Extra Class operator segment of the amateur bands. The distinction here is the Technician Class licensee can't be the control operator where only Extras have privileges.

The station licensee must designate the station control operator. If you have an Extra Class operator in the shack with you, turn the reins over to him and you can operate in the Extra bands.

WHO IS IN CONTROL?

The FCC presumes the station licensee to be the control operator of an amateur station unless documentation to the contrary is in the station records. If someone is sitting at my station and using my call sign, the FCC presumes I am in control. I could prove otherwise if I have a logbook showing another licensee was actually in control at the time.

The rules also say **when the control operator is not the station licensee, the control operator, and the station licensee are equally responsible for the proper operation of the station.** If something is wrong, you're both guilty.

The amateur station control point is the location at which the control operator function is performed. There are three different kinds of control: local, remote and automatic.

Local control is when the control operator is at the control point. You are there, twiddling the knobs on the radio. If you are transmitting on a handheld radio, you are in local control.

Not long ago, I spoke with an amateur using a call sign in the Azores. I recognized his voice, so I asked "Hey Martti, how's the weather in the Azores?" Martti came back, "I don't know. It is cold and snowy here. I am sitting in my living room back home in Finland." Martti was operating his remote station in the Azores over the Internet. That's a pretty amazing bit of technology considering the operator was thousands of miles away from the equipment. The control point was in Finland. My friend in Finland, **operating his station over the Internet - an example of remote control.** He was twiddling the knobs from afar.

Automatic control is when there are no adjustments made to equipment, no spinning of dials or twisting of knobs to change frequency or adjust volume.

Repeaters are an example of automatic control.
The equipment operates automatically, and you don't
have to adjust anything or be in front of it.

**What kind of stations can automatically
retransmit? Auxiliary, repeater and space
stations.** They are all repeaters and repeaters
automatically retransmit.

We will have lots more to say about repeaters later.
Just know for now that **if someone uses a repeater
and violates FCC rules, the control operator of
the originating station is accountable.**

Suppose your brother-in-law is cussing up a blue
streak in the background. If your microphone picks it
up and broadcasts it over the repeater, it is your fault
because you were the control operator of the
originating station. The repeater is on automatic
control. There is no one at the repeater to tell your
brother-in-law to pipe down. You are the one in
control.

BANDS AND FREQUENCIES

Let's begin at the beginning again and talk about some basic theory. **The abbreviation, RF, refers to radio frequency.**

The name for electromagnetic waves that travel through space is radio waves. RF is an electromagnetic wave because it has **two components: an electric and a magnetic field.** You can't see it, so to me, it is just magic. A signal leaves my wire and is picked up on the other side of the world. To quote Samuel Morse's first message, "What hath God wrought?"

A radio wave travels through free space at the speed of light. The approximate velocity is 300,000,000 meters per second.

Radio waves act like alternating current. That is, the direction of flow reverses in cycles, unlike a battery which is direct current flowing only in one direction. **Direct current is the name for current that flows only in one direction. Alternating current is the name for current that reverses direction on a regular basis.**

Frequency is the term describing the number of times per second that an alternating current reverses direction. We measure frequency in cycles per second. The term Hertz is a shortcut for "cycles per second." **The unit of frequency is Hertz.**

The power coming out of an outlet in your house reverses 60 times per second and thus is 60 Hertz – abbreviated 60 Hz. WMAL, our local AM radio station, broadcasts on a frequency of 630,000 Hz or 630 kiloHertz, abbreviated 630 kHz and standing for 630 thousand Hertz. A typical Amateur Radio repeater

might operate on a frequency of 147,015,000 Hz or
147.015 MegaHertz, abbreviated MHz, meaning
147.015 million Hertz.

**The electromagnetic spectrum is commonly
broken down into three groups:
3 to 30 MHz HF (High Frequency)
30 to 300 MHz VHF (Very High Frequency)
300 to 3000 MHz UHF (Ultra High Frequency)**

If we know the wave is traveling at 300 million meters
per second, and the frequency is 144 million cycles per
second, we can calculate how far the wave will travel
in one cycle. **Wavelength is the name for the
distance a radio wave travels during one
complete cycle.**

The higher the frequency, the more often the wave
reverses direction and the less distance it can travel in
a cycle. Therefore, it is said, **the wavelength gets
shorter as the frequency increases.**

**The formula for converting frequency to
wavelength in meters is "wavelength in meters
= 300 divided by frequency in megaHertz."**
For example, if the frequency is 150 MHz, the
wavelength is 300 divided by 150 or 2 meters. The
same formula works in reverse: Divide 300 by the
wavelength in meters to come up with frequency.
300/2 meters = 150 Mhz.

Let's jump now to the concept of a "band." A band is
a group of frequencies. You're no doubt familiar with
the AM radio band that stretches from 550 kHz to
1600 kHz. The 2-meter amateur band goes from 144
– 148 MHz.

**The approximate wavelength property of radio
waves is often used to identify the different
frequency bands.** For example, **146.52MHz is in**

BANDS AND FREQUENCIES

the 2-meter band. (300/146.52 = about 2) The math doesn't work out exactly because 300 divided by 146.52 is not exactly 2 but its close and bands refer to a range of frequencies, so we round off to 2 meters. *Cheat: The other possible answers are wildly off.*

After a while, you will recognize the various wavelengths and the associated frequencies. There aren't that many ham bands to remember.

Here is another example from the test. **The 23 cm band is 300/.23 = 1304 MHz, and 1296 MHz is the closest answer.** The trick there is to remember 23 cm, 23 centimeters is equal to .23 meters.

223.50 MHz is in the 1.25-meter band. 300/223.5 = 1.34. The other answers are nowhere close. Watch your decimal point though because it's easy to make a mistake.

The one exception to the "closest answer" is for 6 meters. **The frequency 52.525 MHz is within the 6-meter band.** 300/6 = 50 and that is the bottom of the 6-meter band. *Cheat: 52.525 MHz is a repeating number.*

After a while, you'll catch on, and when someone refers to the 2-meter band, you'll know they mean 144 to 148 MHz without thinking.

Your Amateur Radio license authorizes you to transmit on certain bands. Not all frequencies are available to amateurs, and not all amateurs can transmit on all amateur frequencies. The higher the class license, the more operating privileges. Most hams have a chart in the shack because it is easier than trying to memorize the restrictions.

For the most part, amateur bands are only available to Amateur Radio operators. However, we share some

bands with other services.   **When an amateur
frequency band is said to be available on a
secondary basis, you must avoid interfering.**
*Hint: We are secondary, and you must defer to the
primary user.*

**If you learn that you are interfering with a
radiolocation service outside of the United
States, you must stop and take steps to
eliminate the harmful interference.** *Hint: The fact
there is a "radio location service" tells you this is a
shared band. Hams are always secondary on shared
bands. If you are interfering, eliminate the harmful
interference by changing frequency or going silent.*

The FCC and ITU dictate certain bands of frequencies
available for amateur operation. **You don't want to
get too close to the edge of the band because:
Your radio might be off frequency
Your transmission is wider than you think
Your radio might drift as it heats up.
All these answers are correct.**
*Hint: There are lots of reasons not to get too close to
the edge. All the choices are correct.*

MODES

The mode is the transmission type. You know about AM and FM modes from your car radio. They are different means of modulating and receiving radio signals. **Modulation refers to combining speech with an RF carrier.** The RF carrier is the base frequency of the transmission. AM is amplitude modulation and FM is frequency modulation. There lots of modes that Amateur Radio operators use.

The following are digital communications modes:
PSK31
Packet
MFSK
All these choices are correct
Hint: Recognize, don't memorize. If you recognize two out of the three, the answer must be all of the above. There is more on digital modes in the chapter about computers.

NTSC is color TV. *Cheat: C for color.*
The widest mode is amateur television because transmitting video requires a great amount of information. **The typical bandwidth of analog fast-scan TV transmissions is about 6 MHz.** That is the equivalent of 40,000 CW signals. *Cheat: Remember TV6*

If you are going to do weak signal work, you need a multi-mode transceiver so you can operate modes other than FM. FM requires a much stronger signal.

FM MODE

In the FM mode (frequency modulation), information is communicated by jiggling the frequency of the transmission resulting in a signal which is relatively wide but clear and high fidelity. The frequency jiggling is called deviation.

Increased deviation makes the signal occupy more bandwidth. The amount of deviation is determined by the amplitude (loudness).

If you are told you are over-deviating, move away from the microphone. Your signal is too wide and might interfere with others.

FM is most commonly used for VHF and UHF voice repeaters.

The approximate bandwidth of a VHF repeater FM phone signal is between 10 and 15 kHz.

Because the signal is wide, it can carry a lot of information. **Frequency modulation (FM) is most commonly used for VHF packet radio transmission.**[5]

AM MODE

In the AM Mode (amplitude modulation), information is communicated by jiggling the amplitude of the radio wave. The audio information varies the power of the radio frequency wave, and the result is the center frequency called the carrier and a sideband on either side of the carrier varying in amplitude in accord with the audio. AM was the phone mode when I was getting started. But AM requires lots of power and heavy duty equipment to handle that power.

---

[5] Packet radio is data transmission. It got that name because the data is transmitted in clumps or packets.

SINGLE SIDEBAND MODE (SSB)

Single sideband sounds like Donald Duck until tuned in properly. But, SSB is a superior mode because it requires less power and takes up less bandwidth. Eventually, SSB won out over AM and today you will only hear occasional AM from boat-anchor enthusiasts[6] – those folks with nostalgia for vintage radios, vacuum tubes and the good-old days.

Single sideband (SSB) is a form of amplitude modulation (AM). Look at the diagram. The bottom waveform is an AM signal. To make SSB,

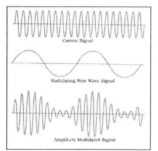

filters in the transmitter strip away the carrier and one of the sidebands leaving only a single side band. The remaining sideband could be either on the upper or lower side of the carrier. You'll hear this referred to as upper sideband or lower sideband. By convention, upper sideband is used on higher frequencies.

Upper sideband (USB) is normally used for 10 meter HF, VHF, and UHF single sideband communications. *Cheat: The answer to remember for the test is "upper." There is no test question with a correct answer of "lower."*

The approximate bandwidth of a single sideband voice signal is 3 kHz. (3000 Hertz) SSB is quite a bit narrower than FM (10-15kHz) or AM (6kHz). *Cheat: Can you remember SSB has 3 letters?*

The primary advantage of single sideband over FM for voice transmissions is SSB signals have a

---

[6] A "boat-anchor" is a heavy old piece of equipment suitable for anchoring a boat.

narrower bandwidth. More signals can fit in a given part of the radio spectrum. Narrower signals also mean you can use narrower filters to reduce noise and interference.

Our early radios had very poor selectivity. That meant you heard many signals, and it was it difficult to pick out the one signal you wanted to hear. **The ability to discriminate between multiple signals is called selectivity.** To get more selectivity, we use filters to reduce the bandwidth of the receiver.

You would use a filter 2400 Hz wide on SSB. That is 2.4 kHz and just slightly narrower than the signal.

SSB voice modulation is most often used for long distance or weak signal contacts on the VHF and UHF bands. It is the primary audio mode on the HF bands as well.

If the voice pitch seems too high or too low, you would use the receiver RIT (Receiver Incremental Tuning) or Clarifier control to fine tune. The voice is off because you are slightly off frequency. The RIT or Clarifier fine-tunes the receive frequency without changing your transmit frequency.

CW MODE (MORSE CODE)

CW, which stands for continuous wave or carrier wave, is the most basic of digital modes. The carrier is turned on and off making the dots and dashes we read as Morse Code. There were several versions of Morse Code, but **International Morse code is used when sending CW in the amateur bands.**

You are not allowed to use codes or ciphers that hide the meaning of a message except when transmitting control commands to space

MODES

stations. Morse Code is not considered a cipher because it is widely used and easily decoded. It, and the other digital modes, such as PSK31, MFSK and packet are not intended to hide the meaning of a message.

The following can be used to transmit CW in the amateur bands:
Straight key
Electronic keyer
Computer keyboard.
All the choices are correct.
Hint: Lots of ways to send CW, so all are correct.
A straight key is a simple on-off switch activated by pumping a lever up and down. An electronic keyer has a lever that moves side-to-side generating a string of dots or dashes depending on whether you push it to the right or left. A keyboard activates the software to send the appropriate code when you push a letter key.

In comparison to SSB and FM, CW emission has the narrowest bandwidth. **The approximate maximum bandwidth required to transmit a CW signal is 150 Hz.** That means 20 CW signals can fit in the same space as one SSB signal.

The typical filter used for CW is 500 Hz wide. The narrower filter cuts down on interference and noise outside the passband.

The advantage of having multiple bandwidth choices is that you can select a bandwidth that matches the mode. If your radio had only one bandwidth, say 6 kHz for listening to AM, you could hear 40 stations sending CW, and that would be terribly confusing. On the other hand, if you only had a 500 Hz filter designed for CW, you wouldn't be able to understand an SSB signal that is 2400 Hz wide.

BANDPLANS

The rules reserve certain parts of each band for different modes. **The 6 meter, 2 meter, and 1.25-meter bands, available to Technician Class operators, have mode restricted sub-bands.** *Cheat: How are you going to remember that? When asking about mode-restricted sub-bands, only one answer mentions 1.25 meters. Recognize "1.25" and pick the answer with that in it.*

Bandplans reserve the lower part of a band for CW. **CW is the only mode permitted in the mode restricted sub-bands at 50.0 to 50.1 MHz and 144.0 to 144.1 MHz.** *Cheat: When asking about mode, the test answer is always either CW or Data. In this case, CW is the only answer offered.*

The emission when operating between 219 and 220 MHZ is Data. *Cheat: The test answer has to be either CW or Data. Look for that in the answer.*

There are also voluntary guidelines beyond the privileges established by the FCC. These are voluntary "band plans."

The national calling frequency on the 70 cm band is 446 MHz. That is the watering hole where everyone goes to look for a contact. *Cheat: When the test asks for a "national calling frequency" you only need to recognize one answer: **446 MHz.***

PROPAGATION

Propagation is the term used to describe the distribution of radio waves. Radio waves travel in straight lines, and unless they bounce off something, communication is limited to what is called "line of sight." The curvature of the Earth blocks radio waves. **The radio horizon is the distance at which radio signals can connect on a direct path.** Buildings, mountains or other physical obstacles can also block radio waves.

When signals bounce off multiple buildings or other objects, it can cause distortion. Just like it is hard to understand music or conversation in a big boomy room with lots of echoes, radio echoes can be distracting. These echoes are "multi-path" distortion.

If the other station reports your signal was strong a moment ago but is now weak or distorted, try moving a few feet or change the direction of your antenna as reflections may be causing multi-path distortion. You may have moved into the shadow of a building or passing truck.

The term, picket fencing, is commonly used to describe the rapid fluttering sound sometimes heard from mobile stations that are moving when transmitting. Picket fencing is a rapid in-and-out flutter usually caused by multi-path distortion.

The part of the atmosphere called the ionosphere enables the propagation of radio signals around the world. The ionosphere is composed of clouds of ions energized by the sun. Signals bounce (refract) off the ionosphere and are reflected back to Earth; a phenomenon called "skip." In some cases, the signals will bounce off the ionosphere back to Earth back to the ionosphere and back to the Earth again for what is called "multi-hop" propagation. Signals can travel around the world in this manner.

The peak of the sunspot cycle makes long distance communication possible on 6 and 10 meters. The peak of the sunspot cycle brings more solar energy to the ionosphere and better propagation.

Generally, the best time for long distance 10-meter band propagation is during daylight hours when there is high sunspot activity. *Cheat: Daylight and Sunspots are good for 6 and 10.*

The 10-meter band is around 28 MHz, and you can operate phone, CW and data modes there with your Technician license.   Please don't think your Technician license limits you to shack-on-a-belt HTs and repeaters. As a Tech, you have limited privileges on HF bands and can experience worldwide communications with a modest station.

Direct UHF signals (not on a repeater) are rarely heard from stations outside your local coverage area because UHF signals are usually not reflected by the ionosphere. VHF and UHF frequencies pass through the ionosphere and out into space. That is why VHF and UHF frequencies are used to communicate with satellites.

VHF and UHF radio signals usually travel somewhat farther than the visual line of sight distance between two stations because the Earth seems less curved to radio waves than to light. *Cheat: That sounds bizarre, so it's an easy test answer to remember.* Air bends the radio wave over the horizon causing the phenomenon.

Signals can also bounce off buildings or bend around a sharp corner. **The term "knife-edge" propagation refers to signals that are partially refracted**

around solid objects exhibiting sharp edges. The term "refracted" is another way of saying "bent."

Radio waves can refract off more than just the ionosphere and buildings. One example is called "tropospheric ducting." **Tropospheric ducting is caused by temperature inversions in the atmosphere.** The radio waves are trapped between layers of different temperature air and bounce along until they squirt out some ways away.

Tropospheric scatter mode is responsible for allowing over the horizon VHF and UHF communications to ranges of approximately 300 miles on a regular basis. *Cheat: Forget the overly-complicated question, if you see "tropospheric scatter" in an answer, it is correct.*

Another propagation example is called "aurora scatter." Signals traveling near the North or South Pole may reflect off the Aurora Borealis, or northern lights. **Signals exhibiting rapid fluctuations of strength and often sounding distorted is a characteristic of signals received via auroral reflection.** On HF, signals traveling over the North Pole from Asiatic Russia will often sound spooky, watery or fluttery. This sound is classic aurora.

When meteors pass through the atmosphere, they leave behind a short-lived trail of ionized gas. Radio waves will bounce off the trail. The 6-meter band can be very active during times of known meteor showers. **The 6-meter band is best suited to communicating via meteor scatter.** *Cheat: "Meteor" has 6 letters, 6 meter = meteor.*

The ionosphere is divided into several layers commonly referred to as the D, E, and F layers. These layers are energized by the sun and exhibit different characteristics depending on the time of day, time of year and the amount of energy coming from the sun.

The E layer is unpredictable and sporadic ("Sporadic E"). **When VHF signals are being received from long distances, signals are possibly being refracted from a sporadic E layer in the atmosphere.**

Occasional strong over-the-horizon signals on the 10, 6, and 2-meter bands are Sporadic E. *Cheat: If "sporadic E" appears in the answer, it is always right.*

Cheat repeat: If you see "tropospheric scatter or "sporadic E" in and answer, it is always right.

UHF works better than VHF from inside buildings because it tends to penetrate the structure. This lesson was brought home on 9/11 when the fire department's VHF radios were deaf inside the Twin Towers. If you operate from inside a building, you could use UHF to get your signal out.

Some VHF/UHF mobile radios allow cross-band repeating. You operate on UHF from inside the building and set up the cross-band repeater in your car outside the building. The cross-band radio receives on UHF and rebroadcasts on VHF, thereby providing a link to the local VHF repeater system.

BE SAFE, STAY SAFE

A fuse protects circuits from current overloads.
A thin metal strip in the fuse overheats and melts,
breaking the connection. **A fuse interrupts the
power in case of an overload**

**Equipment powered from a 120V AC power
circuit should always include a fuse or circuit
breaker on the hot conductor.**

**Don't put a 20 amp fuse in place of a 5 amp fuse
or excessive current could cause a fire.** You lose
protection as this would allow 20 amps to flow in a
circuit only designed to handle 5 amps. The overload
would cause the circuit to overheat.

The following question assumes you are talking about
the plug you stick in a wall socket.
To guard against electrical shock:
Use three-wire cords
Connect to a common safety ground
Use ground fault interrupters.
All the answers are correct.
Hint: There are lots of ways to guard against shock.

The green wire on a plug goes to safety ground.
Hint: Green for ground. That is the lone pin on an
electric plug.

**When putting up a tower, look and stay clear of
overhead wires.** Overhead power wires carry lethal
voltages. If you contact an overhead wire with a
metal antenna support, you will be seriously injured, if
not killed. Look up before you put up.

**What is the minimum safe distance from a power
line? Make sure the antenna can come no closer
than a minimum of 10 feet**. If your pole is 30 feet
long, be sure you are at least 40 feet from the power

line. Measure and put something on the ground, so you don't wander into danger.

Don't attach an antenna to a utility pole because you could contact the high voltage lines.

Use a climbing harness and safety glasses when climbing a tower. The climbing harness will keep you from falling because it attaches you to the tower.

Wear a hard hat and safety glasses below the tower at all times. A tool falling from the top of a tower can cause serious injury.

It is never safe to climb a tower without a helper or observer.   Use the buddy system.

A gin pole is an extension that is used to lift tower sections or antennas above the top of the tower.

Never climb a crank-up tower unless it is fully retracted. Riding a ladder down looks funny in National Lampoon's "Christmas Vacation" but it is very dangerous.

Local electrical codes set grounding requirements. Grounding and house wiring are local electrical code issues. The FCC is not involved.

The best conductor for RF grounding is flat strap. RF travels on the surface, a phenomenon known as "skin effect." The flat strap has the more surface area than wire, so it is more effective for grounding RF.

Proper grounding for a tower is a separate eight-foot rod for each leg bonded to the tower and each other.

SAFETY

For lightning protection, you want ground wires that are short and direct. "Short and direct" makes the maximum power dissipate into the ground. **For lightning protection, sharp bends must be avoided.** Short, straight and direct. Bends impede lightning's direct path to ground.

Ground all the protectors to a common plate which is in turn connected to an external ground. There should be no opportunity for different voltages between different pieces of equipment. Even a slight difference in the resistance to ground can cause a lot of current to flow through your equipment with damaging results.

VHF and UHF radio are non-ionizing radiation. RF radiation differs from ionizing radiation (radioactivity) because RF does not have sufficient energy to cause genetic damage. No evidence yet that RF causes cancer.

But RF exposure can be dangerous. We heat food with RF in a microwave oven, and our bodies can be heated by RF as well.
The factors affecting RF exposure are: Frequency and power level Distance from the antenna Radiation pattern of the antenna. All the test answers are correct.
Hint: Lots of factors affect RF exposure, so all the choices are correct.

Exposure limits vary with frequency because the human body absorbs more RF at some frequencies than others.

50 MHz has the lowest value for Maximum Permissible Exposure Limit. Meaning, 50 MHz heats our body the most. T50 MHz is our 6-meter band.

If you run more than 50 watts at the antenna on VHF frequencies, you are supposed to do an RF exposure evaluation. *Cheat: The answer, both here and above, is "50." 50 watts and 50 MHZ.*
To stay in compliance, you should re-evaluate whenever you change equipment.

An acceptable way to check exposure is using:
FCC Bulletin
Computer modeling
Actual measurements
All the answers are correct.
Hint: Lots of ways to check, so all the choices are correct.

To prevent exposure, try relocating the antenna. Moving further away from the antenna will reduce your exposure.

Duty cycle is the percentage of time the transmitter is transmitting. Duty cycle is a factor because it affects the average exposure.

If the limit is 6 minutes and you are on 3 and off 3, you can be exposed safely for 2 times as much. That would be a 50% duty cycle, and your body has time to cool during the off times. Therefore, you could be exposed safely for twice as long.

A person touching your antenna might get a painful RF burn. RF burns are very painful, and they linger. Stay away from the antenna.

RADIO OPERATION

Now that we've covered basic concepts of frequency, mode and propagation, let's start operating.

A ham shack from the 60's would have included a transmitter and a receiver – two separate boxes. You can see that in my shack picture from 1967 on page 16. Today's radios are **transceivers which combine a transmitter and receiver** in one box. The two automatically track each other with one knob.

If you are transmitting and receiving on the same frequency, it is simplex communication. *Hint: One frequency is simple(x).*

Your microphone has a **PTT (Push To Talk) switch** that switches operation from the receiver side of the transceiver to the transmitter side.

Some microphone connectors include pins for PTT and voltage to power the microphone. Different manufacturers may use different pin assignments or different plugs. If you are buying a microphone, make sure you get the right connector for your rig.

Modern radios will let you **store the frequency in memory channels for quick access.** This is much easier than spinning the dial to get to your favorite frequencies. **You can also enter your frequency by typing it on a keypad or use the VFO knob.** VFO stands for "variable frequency oscillator" and you would be spinning the knob to change frequency.

An oscillator is the circuit that generates a signal at a desired frequency.

Your HT or Handi-Talkie is a form of a transceiver.

The short rubberized antenna that came on top of your HT is called a "rubber duck." **A disadvantage of the "rubber duck" antenna supplied with most handheld radio transceivers is it does not transmit or receive as effectively as a full-sized antenna.** It is handy but too short.

A good reason not to use a "rubber duck" antenna inside your car is that signals can be significantly weaker than when it is outside of the vehicle. The metal car body acts like a shielded cage trapping the signals inside.

Put an antenna in the middle of the roof for in-car use. The middle of the roof provides a uniform reflective surface that won't skew your signal in one direction. A "mag-mount" antenna has a strong magnet that holds in place even at highway speeds. Run the cable through the door jamb and connect it to your HT or mobile radio.

You could increase the low power output from the handheld using an RF power amplifier.
Hint: Increase power with an amplifier.

Crackling static or white noise can be very fatiguing and is guaranteed to drive spouses crazy. **The squelch control mutes the receiver output noise when no signal is present.**

Muting controlled by the presence or absence of an RF signal is called carrier squelch. The squelch deactivates the receiver until the radio senses a carrier signal. *Hint: Tie the words "squelch" and "mute" together.*

There are a dozen or so Q signals in common use. You only need to know these two for the Technician test. Q signals are both shorthand and a universal language. If I tell a Russian there is QRM, I am telling him **"I am receiving interference" QRM.** He doesn't need to speak English, and I don't speak Russian, but the point is understood.

Another example is **I am changing frequency: QSY.**

Contacting as many stations as possible during a specific time is contesting. In a radio contest, you send only the minimum information needed. The idea is to contact as many stations as possible not to chew the fat.

There are several contests every weekend and some hams are fierce competitors. It is possible to work all 50 states or 100 countries in one weekend, qualifying you for a Worked All States (WAS) or DXCC (DX Century Club) award from ARRL, the American Radio Relay League. ARRL calls itself the national association for amateur radio (arrl.org).

"DX" is shorthand for distance. Since my contact with Tom Christian, I have been an avid DXer, chasing new countries. It got so bad one summer, my Mom kicked me out of the basement and told me to play outside. She said my skin was turning blue-green from lack of sunlight.[7]

A Grid Locater is a letter-number assigned to a geographic location. "Maidenhead Grid Squares" divide the world. Someone might not know the location of Fredericksburg, VA but they can find Grid Square FM18. VHF and UHF contests often use Grid locator squares as part of the exchange.[8]

---

[7] Check out my new book "How to Chase, Work & Confirm DX – the Easy Way." It is available from Amazon.

[8] A contest exchange is the information you send and receive to validate the contact. In a VHF/UHF contest it might be a signal report and grid square.

HOW TO START A
CONVERSATION

If you are not using a repeater, you would call CQ. **CQ is a general call for any station.** *(Seek you?)* I can remember operating a demonstration ham radio station at the New York World's Fair in 1964. I called CQ forever while unsuccessfully trying to scare up a contact on 2 meters. It was embarrassing. That was before repeaters. You do not use CQ on a repeater.

On a repeater, in place of CQ, just say your call sign to indicate you are listening. "K4IA listening."

To call another station on the repeater, if you know his call sign, transmit the other station's call sign followed by your call sign. "W3ABC this is K4IA."

Before you call CQ you should:
Listen first
Ask if the frequency is in use
Make sure you are in band
All the answers are correct
Hint: Lots to do before you CQ.

If two stations are on the same frequency and interfere, use common courtesy, but no one has an absolute right to a frequency.

To respond to a CQ, say his call sign followed by your call sign. W3ABC calls "CQ, CQ, CQ. This is W3ABC, Whiskey 3 Alpha Bravo Charlie." I respond, "W3ABC this is K4IA, Kilo 4 India Alpha."
Hint: His call sign first because you want to get his attention.

EXTENDING YOUR RANGE WITH REPEATERS

You can get on the air with an HT for less than $50. You'll be able to talk line-of-sight a couple of miles on simplex and much further through the use of a repeater. **Simplex communication is the term used to describe an amateur station transmitting and receiving on the same frequency. You would use simplex if you can communicate directly without using a repeater.** No need to tie up a repeater when you can communicate directly.

We already discussed the fact that 2-meter (VHF) radio waves do not bounce off the ionosphere. They will bounce off buildings, and if you're high enough or conditions are just right for tropospheric ducting, you might be able to communicate over a few hundred miles. But we need some help for reliable local communication. That's where a repeater comes in handy.

A repeater is a combination of receiver and transmitter that receives a signal and simultaneously retransmits it or "repeats" it to relay over a longer distance.

Since 2-meter communication is line-of-sight, the higher you are the further you can see and the further the radio waves will reach – another way of saying you have extended the radio horizon. A repeater antenna should be on a very high site– usually on top of a tall building, water tower or mountaintop.

Second, the transmitter portion of the repeater transmits with more power than the typical handheld or a radio mounted in your car.
The combination of extra height and power stretch the range out to perhaps 10 miles for an HT or 50 miles

for a full-power mobile radio. Remember our old formula for the area of a circle? A=PiR² A= 3.14 x 10 X 10 = 314 square miles of coverage from your trusty HT or 7,850 square miles from a mobile station. That's the basics of a repeater.

An auxiliary station transmits signals over the air from a remote receive site to a repeater for transmission. It is a repeater for a repeater. **What kind of stations can automatically retransmit? Auxiliary, repeater and space stations.** They are all repeaters and repeaters automatically retransmit.

Usually, a local club builds and maintains the repeater with member dues. That's a good reason to join and support your local club! **Four or more can form a club** and get a **club call sign which is administered by a trustee.** *Hint: Four to form*

Simplex communication is the term used to describe an amateur station transmitting and receiving on the same frequency. A repeater operates in "duplex" mode. Since a repeater is receiving and transmitting at the same time, it makes sense that it cannot do both on the same frequency, or it would interfere with itself. The repeater listens and talks on two different frequencies simultaneously to prevent self-interference.

The common meaning of the term "repeater offset" is the difference between a repeater's transmit and receive frequencies. The most common repeater offset in the 2-meter band is plus or minus 600 kHz. For example, my local repeater receives on a frequency of 147.615 MHz and retransmits on a frequency of 147.015 MHz. To access the repeater, you would tune your receiver to 147.015 so you can hear the repeater and you would set the offset on your radio to transmit +600 kHz so the repeater can hear you.

REPEATERS

A common repeater frequency offset in the 70cm-band is plus or minus 5 MHz. The 70cm-band is much wider than the 2-meter band, and there is plenty of room to spread out, so the repeater offset is much greater.

A way to enable quick access to a favorite frequency on your transceiver is to store the frequency in a memory channel. Your radio will have memory channels, so once you set frequency and offset data in the memory, you do not have to reenter it every time you turn on your radio or switch to a different repeater.

All these frequency allocations and offsets require coordination to prevent interference. **Amateur operators in a local area select a Frequency Coordinator. A Frequency Coordinator recommends transmit/receive channels and other parameters for repeater stations.**

There's another way to prevent interference between repeaters. Many repeaters use a sub-audible tone as a "key" for access. It is sub-audible meaning too low in pitch for you to hear. You program the sub-audible tone in your radio's memory along with the other repeater data.

Continuous Tone Coded Squelch System (CTCSS) is the term used to describe the use of the sub-audible tone transmitted with normal voice audio to open the squelch of a receiver. *Hint: Remember, CTCSS opens the squelch.*

If you're trying to access a repeater and failing, chances are you are you may not have the proper CTCSS, DTMF or audio tone for access. DTMF is the tone generated by your keypad. Some repeaters require a DTMF code. Don't let all those fancy names fool you. They are just another way of saying you may need a special tone to unlock the repeater.

You can learn the local repeater offsets and tones from a published Repeater Directory, by talking to local hams or on the Internet. There are APPS for iPhone and Android that use the GPS on your phone to locate the repeaters around you and provide access information.

While we are on the topic of repeaters, let's talk about the Internet. **An Internet Gateway links a radio to the Internet. The Internet Radio Linking Project (IRLP) connects repeaters to the internet**. Dial in the "phone number" of a distant repeater and your signal will pop out of that repeater – even on the other side of the world.

Voice Over Internet protocol is a method of delivering voice over the Internet. Skype is a use of VoIP.

You can access an IRLP node by using DTMF signals. The touchtone buttons on your microphone are DTMF (Dual Tone Multi-Frequency). Use them to dial the number where you want to go.

You select a specific IRLP node using **the keypad to transmit the IRLP node ID.** If you hook into an IRLP node and dial 5600, your signal will travel over the Internet and pop out on a repeater in London.

You can find lists of active nodes that use VoIP from a repeater directory. There's an APP for that, or you can buy a book. The APP uses location data from the GPS in your smartphone to list repeaters around you.

OPERATING SIMPLEX

Let's get on the air. Assume you have your radio set up and you want to talk to somebody. Listen first and select a clear frequency. You don't want to cause harmful interference to another user. **Harmful interference is defined as that which seriously degrades, obstructs, or repeatedly interrupts a radiocommunication service operating in accordance with the Radio Regulations**.
Hint: Harmful interference = seriously degrades obstructs or repeatedly interrupts.

If your station's transmission unintentionally interferes with another station the proper course of action is to identify properly and move to a different frequency.

It helps if you start off on a frequency where other people commonly listen. These are called "calling frequencies." **The national calling frequency for FM simplex operations in the 70cm-band is 446.000 MHz.** *Cheat: The test only asks one calling frequency so memorize 446.*

If you are looking to establish a contact on HF, you can call CQ. **The meaning of the procedural signal "CQ" is calling any station.** You would say, "CQ CQ CQ this is K4IA calling CQ and listening." We don't ever use CQ on a repeater.

Some folks might recommend a longer CQ sequence, and I won't argue with them, just don't overdo it. If you go on and on and on, whoever's listening will get tired of waiting and tune away. If you don't get a response the first time, call again. It's better to use several short CQ calls than one long one.

Before you call CQ, listen if the frequency is already in use. After listening, you still can't be sure the frequency is not in use because you can't always hear both sides of a conversation. The proper thing to do is ask, "Is this frequency occupied?" before you call CQ.

When responding to a CQ, you should transmit the other station's call sign followed by your call sign. His call goes first to get his attention. Like this: "W3ABC this is K4IA. How copy?" You might also use phonetics, "W3ABC this is Kilo 4 India Alpha." You don't need to say his call in phonetics. I am sure he knows his call. It's your call he might have trouble understanding. Once he acknowledges you, you've made contact and can start your conversation.

Once you establish communications, you're expected to use the minimum power necessary to avoid interfering with other users. **The FCC rules regarding power levels used in the amateur bands require an amateur use the minimum transmitter power necessary to carry out the desired communication.**

You identify every ten minutes and at the end of your communication (conversation). *Hint: Low power and identify every 10 are always the right answers to many questions.*

If you receive a report that your station's transmissions are causing splatter or interference on nearby frequencies you should check your transmitter for off-frequency operation or spurious emissions. *Hint: Splatter or interference are "off."*

EMERGENCY

Amateur radio bills itself as the communication system of last resort: "When all else fails."

If you need to get the attention of a net control station in an emergency, **Begin your transmission by**

OPERATING TIPS AND STRATEGIES

saying "Priority" or "Emergency" followed by your call sign. Use the same protocol on a repeater if you come across an accident and want to summon help.

If there is an emergency net running, **once you have checked into the emergency net, remain on frequency without transmitting until asked.** If you are not handling emergency traffic, stand by and wait to be called to help. Once you have checked in, net control knows you are there and will call on you if needed. In the meantime, stay out of the way.

You can **operate outside of frequency privileges of your license class in situations involving the immediate safety of human life or protection of property**. You can use any band or mode if needed in of an emergency. Those are the FCC rules, so the following answer is still technically true.

When do FFC rules <u>not</u> apply? Never. FCC Rules always apply. Trick question. The rules apply, but the rules change in an emergency.

There are two main groups that organize for emergency communications. **RACES is an organization for emergency and civil defense**. You get certified by taking classes. *Cheat: Civil defense and Certified = C = RACES.*

ARES is also an organization for emergency communications. You voluntarily register your qualifications and equipment. *Cheat: ARES has an R in it and you register.*

Both RACES and ARES may provide communications during emergencies.

Our local club is ARES affiliated and holds a weekly net on the repeater. The purpose of the net is to familiarize operators with net operations and give them a chance to test their equipment and skills. Check into your local ARES net. It will be good training and help you get over your initial "mic fright."

Good emergency traffic handling is passing messages exactly as received.   That is word-for-word, not interpreted, condensed or editorialized. If the message is, "There are 50 people in the Stafford emergency shelter," you don't pass the message as, "There are a ton of people in Stafford."

One way to ensure a voice message is copied correctly is to use standard phonetic alphabet. Using the standard phonetic alphabet is also a good way to identify your callsign.

Here is the word list adopted by the International Telecommunication Union:

A--Alpha	**J**--Juliett	**S**--Sierra
B--Bravo	**K**--Kilo	**T**--Tango
C--Charlie	**L**--Lima	**U**--Uniform
D--Delta	**M**--Mike	**V**--Victor
E--Echo	**N**--November	**W**--Whiskey
F--Foxtrot	**O**--Oscar	**X**--X-ray
G--Golf	**P**--Papa	**Y**--Yankee
H--Hotel	**Q**--Quebec	**Z**--Zulu
I--India	**R**--Romeo	

Messages have a preamble. **"Preamble" refers to the information needed to track the message as it passes through the system.** Another part of a message is the Check.   **"Check" is a count of the number of words in the message**. If the "check" is 27 but you only count 25 words, you missed something.

SATELLITES

In 1957, the Soviet Union launched the first satellite into space. It wasn't much, just a metal ball 2 feet in diameter. Sputnik broadcast a simple Morse Code message back to Earth: di di di dit   di dit. "HI."

A satellite beacon is a transmission from space that contains information about a satellite. The speed of the dits told Sputnik's temperature and pressure. That was **telemetry: the one-way transmissions of measurements**. *Hint: Telemetry is metering.*

Telecommand is a one-way transmission to initiate, modify or terminate functions. *Hint: Telecommand is commanding.* Later satellites would take commands from Earth.

The world thought the friendly little beacon was saying, "Hello." Those of us who know Morse Code knew better. "HI" is the telegraphic equivalent of LOL. Sputnik was laughing at us because the Russians got to space first. Little Sputnik only lasted 21 days but it set off the space race.

More than 70 Amateur Radio satellites have launched. The Amateur Radio satellites act as repeaters in space. **LEO means the satellite is in Low Earth Orbit.** That is between 100 and 1200 miles.

The satellite acts as a repeater. Your signal goes up to the satellite to be rebroadcasted down. In the process, your signal can be heard over thousands of miles – the entire portion of the Earth's surface that can "see" the satellite. **Space stations, repeaters, and auxiliary stations are all repeaters and can automatically retransmit signals.**

You can talk to a satellite as long as your license allows you to operate on the satellite's uplink frequency. The uplink frequency is the frequency you use to transmit, so you need privileges on that band. Your Technician license has you covered for the UHF and VHF frequencies used by satellites.

You can talk to the International Space Station if you have a Technician or above license. Many of the astronauts are hams, and they have ham equipment on board.

Satellites move quickly. **Satellite tracking programs provide maps showing:**
The position of the satellite
Time azimuth and elevation
The Doppler shift.
All choices are correct
Hint: You need a lot of information to track a satellite, so all the choices are correct.

The inputs to a tracking program are the Keplerian elements. These are equations and solutions that track the satellite as it orbits the rotating Earth.

Listening is further complicated by spin fading and Doppler shift. **Spin fading is caused by rotation of the satellite and antennas.** When the satellite spins, the antennas point away from Earth and the signal drops.

Doppler shift is an observed change in frequency caused by the relative motion between the satellite and Earth station. Just like the pitch of a siren rises as it comes toward you and falls as it moves away, the perceived frequency of the satellite's signal will change.

SATELLITES

UHF signals do not reflect off the ionosphere.
The signal goes right through. That is why satellite communication uses high frequencies.

A satellite operating in U/V mode is using an uplink in UHF and a downlink in VHF on the 70cm and 2-meter bands. Think of it as a repeater with an offset that is using two different bands. *Hint: 70 cm is UHF and VHF is 2 meters.*

Transmitter power on the uplink frequency should be the minimum amount of power to complete the contact. The same rule as always, but it is especially important, so you don't overload the satellite's input side.

There are satellites that are FM voice repeaters and some that use digital modes. **Digital satellites commonly use FM Packet.** Those are short bursts of digital data. FM is a wide mode that can accommodate a lot of data.

COMPUTERS

After the invention of SSB and repeaters, the next great breakthrough in Amateur Radio has been the integration of computers and radios. One use is to connect the radio with a logging program. The computer reads your frequency and mode and puts it in the log for you. The picture on the back cover is me operating Field Day.[9] You can see the notebook computer figures prominently.

Another use of computers is digital communications. **PSK31, Packet, and MFSK are all digital communications.** Digital modes work because **your computer sound card can send sound to the microphone and convert received audio in digital form** for display on the screen.

You can send packet by connecting a TNC, terminal node controller, between your computer and radio where it acts as a modem to convert letters and numbers into tones and tones back to letters and numbers.

Packet is a digital system that sends bursts of data. **Included in a packet transmission are a:**
Header with the call sign of the station to which the information is sent
Checksum which permits error detection
Automatic repeat request in case of error
All the answers are correct
Hint: Packet sends a lot of information, so all the answers are correct.

---

[9] Field Day is an annual exercise the last full weekend in June. Hams go to the field and operate with emergency power seeing how many contacts they can make. My personal goal is 1000 contacts in 24 hours. It is a combination of club picnic, campout and operating. Check with your local club. They are bound to have something planned and will welcome you.

COMPUTERS

An ARQ transmission system detects errors and sends a request to retransmit. (Automatic Repeat Query). If the numbers don't add up, the receiving station automatically asks for a repeat.

The error rates will increase if the signal is arriving on multi-path (meaning the signal has bounced off several different places and has therefore become several different signals arriving at slightly different times). The computer will dutifully keep trying until it gets it right.

APRS is another use for packet. **APRS means Automatic Packet Reporting System. APRS works in conjunction with a map showing the location of stations.**

Data to the transmitter for automatic position reports is supplied by a GPS Global Positioning System receiver. The GPS sends your location to a TNC (Terminal Node Controller) connected to the radio, and the radio broadcasts your position over a packet network. Others can receive the information and read your location from a computer generated map.

The packet network is made up of digipeaters, digital repeaters. **A digipeater operates on automatic control.** It is a repeater.

Your computer sound card also generates **PSK, which stands for Phase Shift Keying. PSK31 is a low rate (slow) digital mode.** A PSK signal will trot across your screen about as fast as most of us can type. Your computer can decode PSK signals that are barely audible. As a Technician, you have data privileges on 10 meters and can operate worldwide PSK there.

RADIO DESIGN

The presence of a signal is determined by the receiver's sensitivity.   A receiver's selectivity is the ability to discriminate between signals. These two goals conflict as the first part of the receiver is designed to be sensitive to a wide range of frequencies (sensitivity). Then, you need to narrow what you hear (selectivity).

To accomplish selectivity, you first convert the incoming signal to an intermediate frequency using a mixer. Then, filtering stages are designed to operate on that one intermediate frequency and provide selectivity.

It all starts with an **oscillator – the circuit that generates a signal at a desired frequency.**

Then, you **convert a signal from one frequency to another with a mixer.**

On the transmit side, **you convert a low powered 28 MHz exciter to 222 MHz output using a transverter**. A transverter is a type of mixer. It is a "transverter" because it also transmits. The keywords here are "exciter and output." That tells us you are converting a transmitter, thus a transverter.
Cheat: If you see transverter in an answer, it is always right.

Ignition noise is a pop-pop sound coming from the spark system of a car**. Ignition noise might go away if you turn on the noise blanker**.

Another receiver circuit is called **AGC – Automatic Gain Control keeps the received audio relatively constant**. So you don't blow out your ears tuning across a loud station.

RADIO DESIGN

If your receiver needs a little help with weak signals, **you would put an RF preamplifier between the antenna and receiver.** It is an amplifier that is "pre" or before the radio and boosts the signal.

A regulated power supply prevents voltage fluctuations from reaching sensitive circuits. If the voltage fluctuates, it can distort your signal; your frequency may warble, or your radio may shut down.

You want a stout regulated power supply that can handle the load with some room to spare. Keep this in mind when you buy a power supply for your desktop radio. A robust supply designed to supply 35 amps will only cost about $30 more than a minimal 23 amp version, but you'll never have to worry about power when you get a bigger radio. I like volt and amp meters on mine so I can see what is going on.

Stow your power supply away from the radio so noise and hum from the supply doesn't get in your signal. I keep mine under the desk.

CLEAN UP THE SIGNAL

You'll know a bad signal when you hear it. You can tell from the chirpy, raspy sounding CW or over-driven, over-processed or garbled audio.

If your audio is distorted or unintelligible, the problem could be caused by:
Location
Low batteries
Being off frequency.
All choices are correct
Your location could be blocking your signal or causing multi-path. Low batteries cause your transmitter to become non-linear (distorted). If you are off frequency on FM, you will sound distorted.

Too high a microphone gain and the output signal may become distorted. Don't overdrive!

On FM, the amount of deviation is determined by the amplitude (loudness). **If you are told you are over-deviating, move away from the microphone.** Your FM signal might interfere with other stations if your **microphone gain is too high causing over-deviation.** Your signal is too wide.

A variable high-pitched whine on the audio is noise on the vehicle's electrical system being transmitted. More specifically, **a high-pitched whine that varies with speed is usually the alternator of your car.**

Since the noise is in your transmitter, you won't hear it, but the person you're talking to might comment. The best cure is to connect your radio directly to the battery. Two things happen. First, you have less chance of being tied to a circuit with noise and second; the battery acts like a huge filter capacitor.

CLEAN UP THE SIGNAL

There is nothing wrong with asking for or giving a critical signal report. Everyone wants to sound good. No one wants to be a "lid."[10]

Sometimes RF can get into the audio and be rebroadcast. That is RF feedback. **A symptom of RF feedback is reports of garbled, distorted or unintelligible transmissions**. Poor grounding or being close to your antenna can cause feedback.

Shielded wire prevents coupling of unwanted signals to or from the wire.

RF can get into your microphone cable, power cable or cables connected to the computer. **To cure distorted audio caused by RF current on the shield of a cable use a ferrite choke.** Ferrite is a mixture of ceramic and metal that greatly increases the effectiveness of a coil resisting the passage of RF. A few turns of your cable through a ferrite choke may solve your problem.

To reduce or eliminate interference to a nearby telephone put an RF filter on the telephone. The phone line is acting as an antenna, and you need to choke off the RF with an RF filter. *Hint: Filter RF with a filter.*

The first step to resolve cable TV interference is to be sure all the connectors are installed properly. Loose connectors break the shield and allow interference to enter or exit. Tighten them down and make sure any cable that is not connected has a terminating plug – a small cap that screws on the unused end of a TV cable.

---

[10] Lid is an old telegraph expression referring to a poor operator. It is still used.

**Radio frequency interference can be caused by
Fundamental overload
Harmonics
Spurious emissions
All choices are correct**
Hint: Lots of things cause interference. All are correct

**An AM/FM radio might receive Amateur Radio
transmissions because the receiver is unable to
reject strong signals outside the AM or FM band.**
That is an example of fundamental overload. You are
just too strong and overwhelm the other radio.

Harmonic emissions are a by-product of amplifiers.
Even with good design, some harmonic radiation may
escape. If you are operating on 28 MHZ, you have a
weak signal also at 56 MHZ. **To reduce harmonic
emissions install a filter between the transmitter
and antenna**. That stops the harmonics before they
get to the antenna and radiate. *Hint: We are
reducing emissions, so the filter goes between the
transmitter and antenna.*

Spurious emissions are trash outside your intended
signal. A defective radio, low power supply voltage,
over-deviation or overdriving an amplifier can cause
spurious emissions.

**To correct radio frequency interference:
Use low and high pass filter,
Band reject and band pass filters,
Ferrite chokes.
All choices are correct.**
*Hint: Recognize the answer has lots of filter solutions.
Don't try to memorize every component.*

A low-pass filter chokes off everything above its
designed frequency (it passes the lows). A high-pass
filter chokes off everything below its designed
frequency (it passes the highs). A band-reject filter

CLEAN UP THE SIGNAL

blocks a particular range of frequencies and a band-pass filter only accepts a certain range. You pick the type of filter to match the problem.

To reduce the overload of a non-amateur radio or TV, block the amateur signal with a filter on the antenna input of the receiver. *Hint: YThe question asks about overload on a non-amateur radio or TV. That is a receiver, so the filter needs to go on the input of the receiver to block the amateur signal.*

If a neighbor complains you are interfering with their radio or TV, make sure your station is not causing interference to your own radio or television. First, check your gear. If you are clean at home, the problem is your neighbor's. Once they see your antenna, you become a target. Hams often get blamed even when not transmitting.

Part 15 devices are unlicensed and may emit low power radio signals on frequencies used by licensed services. Modems, cordless phones, weather stations, wireless printers, audio and video equipment are all called Part 15 devices. Read the label on one. The FCC has authorized such devices with the understanding their owner must accept interference from a licensed service (you) but must not interfere with a licensed service (you again). I don't know how you tell the neighbor to get rid of his plasma TV (a notorious Part 15 violator), but you are in your rights to do so.

If something in a neighbor's home is causing harmful interference to your amateur station, Work with your neighbor Check your station Politely inform. All the choices are correct.
Hint: There are lots of suggestions, so all the choices are correct.

ELECTRONIC THEORY

Let's talk about some of the parts we "rescued" out of the old chassis behind the TV repair store.

The electrical part that **resists the flow of current is called a resistor. Resistance is measured in ohms.** Resistors have color-coded bands that tell you their value.

Volume controls are potentiometers, variable resistors.

Inductors are composed of a coil of wire. Inductors store energy in a magnetic field.

The ability to store energy in a magnetic field is called inductance.

The basic unit of inductance is the Henry. The more inductance, the more henries. *Hint: Tie magnetic field, coils, henries and inductance together in your mind.*

Impedance is the measure of opposition to AC current flow. The higher the frequency, the more a given Henry impedes the flow. **The impedance is measured in ohms** (just like DC). If you know the value in Henries, you can calculate the amount of impedance for a given frequency. We'll leave that math for the Extra license class.

Two or more conductive surfaces separated by an insulator, make up a capacitor. The ability to store energy in an electrical field is called capacitance.

The electrical component that stores energy in an electrical field is called a capacitor.

ELECTRONIC THEORY

The basic unit of capacitance is called a farad.
Hint: Tie electrical field, farads, and capacitors in your mind.

The higher the frequency, the less impedance a capacitor provides. It is the opposite of the inductor.

A simple resonant or tuned circuit is an inductor and capacitor connected in series or parallel to form a filter.

If an Ohmmeter indicates low resistance and then increasing resistance with time, the circuit contains a large capacitor. It is charging up.

When a power supply is turned off, you might still receive a shock from stored charge in a large capacitor.

Capacitors were a whole lot more fun than inductors. We'd get a big capacitor out of a TV power supply, charge it up and then toss the electrostatic bomb to some poor unsuspecting classmate. Of course, they would catch it, and get a nasty jolt. We were way beyond handshake-buzzer mischief. If you did what we did today, you would probably be arrested.

Those old TVs ran on vacuum tubes, but we did plenty of experimenting with transistors. A transistor is sometimes called a semiconductor. **A component made of three layers of semiconductor material is a transistor.**

The three layers are connected to three electrodes emitter, base, and collector.

Transistors are components capable of using voltage or current to control current flow.

A transistor can be used as a switch or an amplifier.

Electronic components that can amplify signals are transistors. The term that describes a transistor's ability to amplify a signal is called gain.

An FET is a special kind of transistor called a Field Effect Transistor. An FET has a source, gate, and drain. *Hint: Just remember a field has a gate and a drain.*

The component that allows current to flow in only one direction is a diode. Diodes convert Alternating Current to Direct Current, and that is their job in a power supply.

Diodes have electrodes called an anode and cathode. The anode is the positive side and the cathode is negative. *Hint: Remember "Annie and Cathy."*

The cathode end of a diode is identified with a stripe. Diodes have polarity and must be installed in the right direction to convert AC to DC in the right direction.

VOLTS, OHMS, AMPERES & POWER – OHM'S LAW

Ohm's law describes the relationship between voltage, amperage and resistance. The math involved is just multiplication or division and is certainly within the grasp of any high school student. The trick is to know which one to use. Here is how you sort them out.

Voltage represents electromotive force. Think of Voltage as pressure in the line. All that force can just sit there as it does with an electrical outlet. With nothing plugged in, the 120 volts is there, but not doing anything.

Amperes represent the current or amount of electricity flowing through the circuit. When you turn on a light, current flows through the wire to power the bulb. Now, that voltage is pushing the current through the load (light bulb).

Resistance also affects the flow. The filament in the light bulb is resisting the flow and heating up as a result. Even a copper wire has some resistance. Remember the over-heating wire from the arc?

Power is the amount of work that can be done by the pressure and flow.

You can intuitively feel there is a relation between all four:
More pressure = more current flow.
More resistance = less current flow.
More current or more pressure = more power.
Can these be related mathematically? Georg Ohm thought so and came up with his famous formula known as Ohm's Law.

Ohm's Law: Voltage = Amperes times Resistance. For this amazing feat, he got to name the unit of resistance after himself (Ohm).

E symbolizes voltage (Electromotive Force).
I is Amps or Amperage (flow).
R is Resistance.

Memorize that: Voltag**E**, **R**esistance, and the other one, Amps, **I.**

So the formula for Ohm's Law looks like E=IR.
If you know two of the variables, you can calculate the third. The easy way is to use the magic circle and draw it as the Eagle flies over the Indian and the Rock.

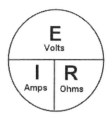

Put your thumb on the value you are solving for, and the formula is what's left. For example, if you are solving for R, cover up the R and the answer is E/I, voltage divided by amperage. If you are solving for E, cover it up and the answer is I x R, amps times ohms.

The questions in the pool that ask you to use OHM's Law are in the Quick Summary starting at question T5D04 (Page 124). They are all variations on the same question. You will get one on your test. It should be easy if you keep calm, use the magic circle and watch your decimal point. Go there now and solve them for practice.

POWER IN WATTS

The rate at which energy is used is called power. Electrical power is measured in Watts.

The formula to calculate electrical power in a DC circuit is Power equals Voltage (E) multiplied by Current (I). It is easy as PIE, P=IE

We have another magic circle to help us.

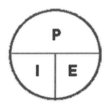

Cover the answer and solve using the two known amounts.

A 100 watt light bulb plugged into a 120-volt circuit will draw 100/120 = .8 amps. A 100-watt bulb in a 12-volt circuit would require 100/12= 8 amps. Higher voltage requires less current to get the job done.

The pool questions are in the Quick Summary starting at question T5A02 (Page 125). Cover the answer and practice. Don't panic. Take your time.

HOW TO DRAW A RADIO

The plans for the guts of electronic equipment are drawn out in a schematic diagram. Schematic diagrams are road maps that show how the components are connected.

Schematic symbols are standardized representations of electrical components. The symbols in an electric circuit diagram represent different electrical components.

The question pool has three schematic diagrams. You will be given only one of them as part of your test package. The test asks you to identify some of the components. The only components you need to identify are in the following discussion. Don't fret over the other components.

The symbols are fair representations of the actual components. Think about the design or function of a part and the right answer will be apparent.

The following questions ask you to look at a schematic diagram and name a few of the components.

Look at Figure T-3 on the next page.

Component 3 is a variable inductor. An inductor is a coil. You can change the value of a variable inductor. One method is by tapping into a different turn. *Hint: The symbol looks like a coil. Imagine the arrow sliding up and down to change the value.*

HOW TO DRAW A RADIO

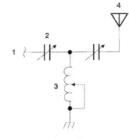

Figure T-3

Component 4 is an antenna. *Hint: It looks like a funnel to collect radio waves.*

A tuned circuit is composed of an inductor and a capacitor. In the case of figure T-3, we have two variable capacitors and a coil to form a tuned circuit. **A simple resonant or tuned circuit is an inductor and capacitor connected in series or parallel that form a filter.** *Cheat: Ditch the overly-complicated question. Just know an inductor and a capacitor together form a tuned circuit or filter.*

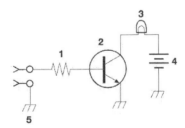

Figure T-1

Looking at Figure T-1 above, **Component 1 is a resistor**. *Hint: Imagine the resistance if you had to run a zigzag line like that.*

Component 2 is a transistor. Review: **A transistor's three leads are: emitter, base, and collector.** The function of the transistor is **to control**

**the flow of current. A transistor can be an
amplifier or a switch.**

A switch connects or disconnects circuits.
The transistor looks like it is a switch in this circuit. A
small change in the voltage applied through the
resistor will cause the transistor to conduct and light
the lamp.

Figure 3, a lamp, looks like a lamp.

Component 4 is a battery with long and short lines
representing the cells. If there were only two lines,
you would be looking at a capacitor.

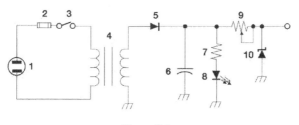

Figure T-2

In Figure T-2, above, **Component 3 is a switch**. The
one in the diagram is **single-pole single-throw**.
Hint: One wire in and one wire out

Component 4 is a transformer. See the sets of
windings (coils). The two coils are coupled together
through mutual inductance. One side induces current
and voltage on the other side. The ratio of the
number of windings determines the step-up or step-
down result.

HOW TO DRAW A RADIO

A transformer is used to change 120v AC house current to lower voltage for other uses. The little cube you plug in to charge your cell phone is a power supply with a step-down transformer.

Component 6 is a capacitor. A capacitor has two sides, and so does its symbol. If there are more than two lines in the drawing, you are looking at the symbol for a battery.

A light emitting diode is commonly used as a visual indicator. The buttons that light up on your TV remote are LEDs. **LED means light emitting diode. Component 8 is a light emitting diode**. *Hint: The arrows are light coming out of the LED.*

Component 9 is a variable resistor. The zigzag line is a resistor; the arrow makes it variable.

Some other components that aren't in the diagrams:

A meter is used to display signal strength (and a lot of other things).

A relay is a switch controlled by an electromagnet. For example, the ignition switch in your car could never handle the tremendous surge of power needed to start the car engine when you turn the key. A small amount of power supplied by the ignition switch closes a relay with big contacts that pass the power from the battery to your starter motor.

The type of circuit that controls the voltage from a power supply is called a regulator. *Hint: A regulator regulates.*

An integrated circuit combines several semiconductors in one package. It integrates (combines) several circuits on one component, sometimes called a "chip."

DECIBELS

Decibels are logarithmic (Power of 10) and relate to relative change. **Going from 20 to 200 watts is 10 times or 10 dBs.** Going from 200 watts to 2000 watts is another 10 db.

Since we are talking about logarithms, the change is not linear. **Going from 5 watts to 10 watts is 3 dB**. Doubling is 3 dB. It doesn't matter where you start, if you double, it is 3 dB, and if you double again, it is 6dB.

If you halve the power it is -3dB. Halve it again and it is -6dB. **A power decrease from 12 watts to 3 watts is -6dB.** Minus half and minus half again

An example of decibels is to compare the results of a better antenna versus an amplifier. A good 2-meter directional antenna might have a gain of 9dB. Each 3dB is a double, so that is double, double and double = 8 times the power. To get the same result for your 50-watt desktop radio, you would have to buy a 400-watt amplifier. The antenna is a lot cheaper and will give you gain on the receiver side as well.

Coax feed line has losses that add up quickly on high frequencies. Check the specs when you chose coax. If it has 3 dB of loss over 100 feet, you will lose half your power in 100 feet of cable.

MOVING DECIMALS

We are usually dealing with very big or very small numbers so it is convenient to leave out all the zeroes and use a new term.

A kilovolt would be one thousand volts. Kilo is thousand. Mega is million. Giga is thousand million.

1,500,000 hertz is the same as 1500 KHz. Kilo means thousand so 1500 KHz is 1500 thousand hertz or 1,500,000 hertz or 1.5 Megahertz (MHz). They all say the same thing.

2425 MHz would be the same thing as 2.425 Ghz. Gigahertz. A Gigahertz is a thousand million hertz. **28,400 KHz would be the same as 28.400 MHz**. *Hint: When going to a larger identifier, put in a comma and then change it to a period.*

For really small numbers we use, milli (1/1000) and micro (1/1,000,000.) and pico (1/1,000,000,000)

One **microvolt is one one-millionth of a volt.**

A 3000 milliampere current would be = 3 amperes. 3000 /1000 *Hint: Put a comma in 3000 and change it to a period.*

500 milliwatts would be 0.5 watts 500/1000

1.5 amperes would be 1,500 milliamperes.
Hint: How many thousandths? 1.5 X 1000 =

Picofarads are really tiny: one millionth of one millionth of a farad. So **1,000,000 picofarads =** 1 millionth of a farad or **1 microfarad** I don't blame you if you have to memorize this one. It is a brain teaser.

BUILDING EQUIPMENT AND MEASURING VALUES

Radio and electronic soldering use rosin core solder. Acid core plumber's solder would eat away the connection.

A cold solder joint is grainy and dull. You heat the joint and flow the solder onto the connection. Don't drip hot solder on a cold connection. If you see a cold solder joint, it means there was not enough heat or the part was moved while the solder was cooling. Reheat the connection. Cold solder joints are poor connections both mechanically and electrically.

Multimeters measure voltage and resistance. Everyone needs a couple of multimeters. They are very handy and some cost under $10. The cheap ones are not laboratory grade, but they will give accurate enough readings for most uses.

If you're measuring resistance, be sure the circuit is not powered. A variation on the same question: **you'll damage the meter if you try to measure voltage when using the resistance setting.**

The resistance setting is very delicate and running voltage through it will pin the needle or let the smoke out of your meter. That is a ham-speak meaning you fried it. If you are lucky, your meter had a fuse that blew before being damaged.

ANTENNAS - PUT SOME FIRE IN YOUR WIRE

POLARIZATION

Radiation is strongest broadside to the antenna.
Whichever way the antenna is mounted or facing,
most of the energy will be broadside to the wire and
not off the ends.

**The orientation of the electrical field describes a
radio wave's polarization.**

**A simple dipole mounted parallel to the Earth's
surface is horizontally polarized**. The wire is
horizontal, and the radiation is strongest broadside, so
the radiation is also horizontal. **The antenna
polarization used for CW and SSB is usually
horizontal.**

**A vertical antenna has an electrical field
perpendicular to the Earth.** Vertical antennas also
radiate broadside, and since the antenna is vertical,
they are vertically polarized. Mobile FM uses vertical
polarization because that is the way an antenna fits on
a car and you hold an HT: antenna up-and-down.

**Mounting the antenna in the center of a vehicle
roof provides the most uniform radiation
pattern**. Mounting your antenna on the corner of a
car can distort the pattern making it favor certain
directions. Since you are moving, the results are
unpredictable. If you have no choice, do what you
can. My antenna is on a clip hooked to a side window,
and it works fine.

**If you are using different polarizations, signals
are significantly weaker.** Don't hold your rubber
ducky sideways as that will make the antenna

horizontally polarized. If you are talking on a vertically polarized repeater, your signal could be as much as 100 times weaker.

When a signal bounces off something, it becomes wobbly like a tumbling hula hoop. It becomes elliptically polarized (random both ways), and it doesn't matter how your antenna is set up. **Skip signals are elliptically polarized, and you can use either a vertically or horizontally polarized antenna**.

Irregular fading can be caused by the random combining of signals arriving from different paths. Signals arriving at slightly different times or different polarizations because they bounced off different objects are classic multi-path.

ANTENNA LENGTHS

I told the salesman I needed some antenna wire. He asked me, "How long do you need it?" I thought that was an odd question so I answered, "I'm building an antenna, so I guess I'll need it for a long time." He coax-ed me out the door.

How long should an antenna be? The classic and most fundamental antenna is a one-half wire wavelength long fed in the middle with coax. If you think about the current flowing in the wire, it will start out high in the middle at the feed point and by the time it gets to the end, the cycle will be ready to reverse. Nothing is left over, and nothing is reflected back.

A half-wavelength antenna is half a wavelength long. The test question asks, **"How long is a 6-meter half-wave dipole?"** Six meters is a full wavelength so 6 meters / 2 = 3 meters long for a half wave. That is a little over 9 feet x 12 = **112 inches is the closest answer.**

ANTENNAS

Many verticals are quarter-wave, and the ground or the vehicle body provide the other half of the antenna. **A quarter-wave for 146 MHz is 19 inches**. If you know the frequency, the magic number is 234. Divide 234 by the frequency and you get the length of a quarter-wave in feet. 234/146 = 1.6 feet. 1.6 feet X 12 = 19.2 inches.

To make an antenna resonant at a higher frequency, you would shorten it. Higher frequency = shorter wavelength. If you set up your antenna and your antenna analyzer shows it to be resonant at a lower frequency than you expected, you would shorten it to bring to the resonant point.

A 5/8 wave antenna offers a lower angle of radiation and more gain than a 1/4 wave. The slightly longer length causes the signal to squash down closer to the horizon. You will see 5/8 wave antennas offered for car installation.

If your calculations tell you the antenna needs to be too long for your installation (an 80-meter dipole is 133 feet long), you can add an inductor to shorten the wire. **A type of loading can be an inductor to make an antenna electrically longer.** You won't have this problem on VHF/UHF because the antenna sizes are quite manageable.

DIRECTIONAL ANTENNAS

Yagis, Quads, and Dishes are all directional antennas. They concentrate the signal in one direction and reject signals to the back and side of the antenna. **A beam antenna concentrates signals in one direction**. We call Yagis "beams" because they beam the signal in one direction.
The gain of an antenna is **the increase in signal strength in a specific direction compared to a**

reference antenna. A directional antenna has gain on both transmit and receive. That is an important consideration. In our discussion of decibels, we found 9dB gain from a beam antenna was the same as adding a 400-watt amplifier. Antennas are reciprocal meaning the antenna also gives 9dB of gain on receive but the amplifier does not. Spend your money on antennas before amplifiers.

Directional antennas could locate the source of interference or jamming with radio direction finding. You get a bearing on the strongest direction from two locations. The target is where the bearings cross on a map. **A directional antenna is also useful for a hidden transmitter hunt**. This sport is also called fox-hunting. You try to find a small hidden transmitter. There are national and international fox-hunting competitions, but the fox doesn't get hurt.

When using a directional antenna, you might be able to access a distant repeater if buildings or obstructions are blocking the direct line of sight path by finding a path that reflects signals to the repeater. *Hint: You point your antenna to bounce around the obstruction.*

COAXIAL CABLES

Coaxial cable has a center conductor, a layer of insulation, then a braided shield all covered with a tough outer cover.

A common use of coaxial cable is carrying RF signals between a radio and antenna. The line carrying RF to the antenna is sometimes called feed line.

Coax is favored over other feed line because coax is easy to use and requires few special installation considerations. Coax is very versatile.

ANTENNAS

You can run it anywhere including next to metal such as aluminum siding. If you have extra, coil it up on the floor. You can install coax outdoors or indoors, and bury certain kinds.

The characteristic impedance most often used for coax is 50 Ohms. It matches the normal impedance of a dipole installed over the ground.

When increasing frequency, coaxial cable loss increases. When choosing coax, check the loss attenuation, measured in dB per hundred feet at various frequencies. You don't want to lose your signal in the feed line. 3 dB means half your power is converted to heat and lost!

Larger RG-8 cable will have less loss than smaller RG-58. *Hint: Telling which cable is larger gives away the answer. Larger cable (bigger pipe) means less loss.*

The lowest loss at VHF would be air-insulated hard line. It is "hard line" because it is very rigid. That makes it hard to bend.

Air core coax has the lowest loss at high frequencies, but the disadvantage is that it requires special techniques to avoid water absorption. Commercial installations use hard line but most hams will avoid it. *Hint: Water absorption is always bad.*

The most common cause for the failure of coaxial cable is moisture contamination. Water gets in, and the braided shield acts as a wick. The leak can come from the connectors or wildlife. Squirrels love to chew on coax, and they gnaw through the outer jacket exposing the braid. I inspect my coax every couple of years and every time the antenna comes down. I always find some squirrel damage.

A PL 259 connector is commonly used on HF frequencies. The PL 259 is a type of screw-on plug that has been around since the 1930's. They are not waterproof, but they are relatively easy to solder to coax cable.

Frequencies above 400MHz use a Type N connector. The type N is a different plug design that is waterproof. It is harder to solder to coax.

Seal the connectors to prevent an increase in feed line loss (from moisture). Minimally, use good quality electrical tape rated for UV (sun) exposure. Other sealing products include self-amalgamating tape that forms a flexible waterproof shield over the connector.

Coax outer jacket should be resistant to ultraviolet light because ultraviolet light from the sun can damage the jacket and allow water to enter. Baking in the sun for years can destroy the jacket causing it to crack or peel. Good quality coax from name-brand manufacturers can last for decades if the squirrels don't chew through it first. It is worth inspecting your coax for damage every couple of years. The loss of signal will be so gradual you will not notice it otherwise.

ANTENNA ANALYZERS AND SWR

When we weren't busy trying to make gunpowder or some other mischief, we would put up antennas. I don't think we knew what an antenna analyzer was, and the concept of SWR was also a bit foreign. We just followed the formulas and hung wire in the air. For the most part it worked, and we made contacts. What you don't know can't hurt you.

ANTENNAS

Old tube rigs had a wide-range matching system that would pretty much tune up into any antenna, so you just didn't worry. Modern solid state rigs are much more sensitive. They want to see 50 Ohm's and will reduce the forward power if too much power is reflected back due a mismatch.

We use an antenna analyzer to determine if the antenna is resonant. By resonant, we mean how well the antenna matches to the frequency. .

You can also measure proper match with a directional wattmeter. It shows the forward and reflected power. You want to minimize the reflected power.

The standing wave ratio (SWR) is a measure of how well the load is matched. Reflected power is compared to forward power. Obviously, less reflected power is better. **You want a low SWR to allow the efficient transfer of power and reduce losses.**

You connect an SWR meter in series between the transmitter and the antenna. It is measuring the forward and reverse power coming out of and into the transmitter and antenna.

A perfect match is an SWR of 1 to 1. Don't get too crazy chasing the elusive 1:1. Often, that is just not achievable, and the difference in signal is not significant. Most of us would be thrilled with 1.5 to 1 and aren't concerned until the radio starts to reduce power.

Most radios will automatically reduce power if they sense an SWR above 2 to 1. If you don't have a good match, power is reflected back down from the antenna. The radio has a built-in defense mechanism.

An SWR of 4:1 would indicate an impedance mismatch.

You can use an antenna tuner to help match the antenna system impedance to the transceiver.
An antenna tuner is a combination of variable capacitors and inductors to make up for the mismatch. If you have a mismatch, it is best fixed at the antenna. Applying a band-aid on the transmitter won't stop the losses that occur in coax when the SWR is too high.

If you have erratic changes in SWR readings, you probably have a loose connection in the antenna or feed line. Make a chart of your SWR readings so you can compare over time and see if anything is changing in your system. If your SWR starts climbing, you may have moisture in the coax.

Power lost in the feed line is converted to heat.
You want signal, not heat.

A dummy load prevents radiation of signals when making tests. We use a dummy load when we don't want our signal to go out over the air.

A dummy load consists of a non-inductive resistor and heat sink. The resistor absorbs the transmitter current, and the heat sink dissipates the heat from power burned up in the resistor.

BATTERIES

**Rechargeable batteries include
Nickel-metal hydride
Lithium-ion
Lead-acid gel cells.
All choices are correct.**
Hint: There are many kinds of rechargeable batteries, so all are correct.

Carbon Zinc are non-rechargeable batteries.

A mobile transceiver requires about 12 volts.
Mobile transceivers use the same voltage as a car battery which is a nominal 12 volts.

You can recharge a 12-volt lead-acid battery if the power is out by connecting it in parallel with a vehicle's battery and running the engine. A car can run on a full tank of gas for about 72 hours. If you ran the car a few hours a day to charge batteries, it could last for weeks. Consider your car as a source of energy in case of a power outage.

If you try to charge or discharge a lead-acid battery too quickly, it can overheat and give off flammable gas or explode. A hazard presented by a conventional 12-volt storage battery is that explosive gas can collect if not properly vented. Take precautions if you plan to keep batteries indoors. You also need to keep them off the concrete. The venting battery acid vapors will destroy concrete.

A safety hazard of 12-volt storage batteries is that shorting the terminals can cause burns, fire or an explosion. Lead acid batteries can deliver a tremendous amount of current that can melt a ring or tool. Take your ring or metal watchband off and keep your tools clear when working around car batteries.

Connect the negative side of the power cable to the battery or engine block ground strap. The negative is the black wire, and it is considered the "ground."

Congratulations! You're done. Now run through the Quick Summary and, when you feel confident you know the answers, try some practice tests online. Go to QRZ.com and click on the Resources tab. Good luck.
73/DX – Buck, K4IA

Here's my HF/VHF/UHF mobile rig installed in a Jaguar XJ6. The radio is in the trunk. A control head is in what used to be the ashtray. The Morse Code paddles are to the left of the gear shift.

This is what I call "the cat's meow."

Amateur Radio Technician Class Quick Summary

In this section, the questions are in regular typeface and the answers are in bold. Learn to recognize the answer.

INTRODUCTION TO AMATEUR RADIO

T1A01[11] What is the purpose of Amateur Radio? **Advancing skills in the technical and communication phases of the radio art**. *Recognize the odd phrase "radio art."*

T1A02 Who administers Amateur Radio in the US? **FCC**

T1A10 FCC Part 97 **defines an amateur station as a station in the Amateur Radio Service.**

T1A03 Which part of the FCC rules govern Amateur Radio? **Part 97**

T1A05 Which of the following is a purpose? **Enhancing international goodwill**.

T1A12 What's a permissible use? **Radio experiments and communicate with others around the world**.

T1C13 License Classes are **Technician, General and Extra**.

T1C08 The normal license term is **10 years**.

T1C09 Grace period of **2 years**.

T1C11 During the grace period **you are NOT allowed to transmit**.

---

[11] This is the question identifier. It indicates the section of the question pool and the question number.

T1D02 You can exchange messages with a US Military station **only during an Armed Forces Day communications Test.**

T1F13 When can the FCC inspect your station and records? **Any time upon request**. .

T1C07 What happens if mail from the FCC is returned**? Returned mail can result in revocation of license.**

T1C10 How soon can you operate? **As soon as operator/station license grant appears in the online FCC database.**

INTERNATIONAL REGULATIONS

T1B01 The ITU is a **United Nations Agency for information and technology issues.**

T1B12 ITU Regulations are not the same everywhere in the world, **A US station operating maritime mobile (on a ship) may be in a different ITU region from the US.**

T1B02 Frequency assignments for some US territories are different from the 50 states **because some US territories are located in ITU regions other than Region 2.**

T1D01 With which countries are FCC-licensed amateur stations prohibited from exchanging communications? **Individual countries have their own rules and may notify ITU they object to communications.**

T1C03 What type of communication is permitted? **Incidental to amateur service and personal nature.**

T1F07 What restrictions apply when a non-licensed person is allowed to speak on your station? This is called "third-party" traffic and the foreign station must be in a country **with a third party agreement.**

T1F11 To which foreign stations do FCC rules authorize third party communications? **Any station whose government permits.**

T1C04 You can operate in a foreign country **if the foreign country authorizes it.**

T1C06 In addition to the places the FCC regulates**, you can transmit from a vessel in international waters if the vessel is registered in the US.**

WHO ARE YOU?

T1F03 When do you transmit your assigned call sign? **Every 10 minutes and at the end of a communication.**

T2A06 **Identify** when making test transmissions.

T2A07 When making a test transmission, **Identification is required at least every 10 minutes and at the end**

T1D11 When can an amateur station transmit without identifying? **Exception allowed for control of model craft**

T8C08 A radio controlled model must have a **label with name, call sign and address.**

T8C07 Maximum power with radio control models? **1 watt**

T1C02 The valid call sign is **W3ABC.**

T12C12 **Any licensed amateur** can select a desired call sign under the vanity call sign rules.

T1C05 **K1XXX is available to Techs**.

T1C01 A call sign with a single letter in both its prefix and suffix is a **special event call sign**.

T2B09 FCC encourages you identify with a **phonetic alphabet.**

T1F12 How many persons required to issue a club license? **Clubs of at least 4.**

T1C14 Who may select a vanity call sign for a club? Club calls are administered by a **Trustee.**

T1F01 "Race Headquarters" is an example of **Tactical call.**

T1F02 But you still must identify with your real call **every 10 minutes and at the end.**

T1F04 You must identify in **English.**

T1F06 **Slash, slant, stroke** are all acceptable self-assigned indicators.

T1F08 Other indicators are optional but you must use a special indicator to operate with your new privileges **while waiting for your upgrade to appear in the FCC database.**

WHO IS IN CONTROL?

T1E01 When are you permitted to transmit without a control operator? **Never.**

T1E02 Who may a licensee designate as a control operator? It must be a **person with a license that appears in the FCC database.**

T1E03 Who designates the control operator? **Station licensee.**

T1E07 When the operator is not the station licensee, who is responsible? **Operator and station licensee both.**

T1E04 What determines the transmitting privileges? The class of license held by the **control operator.**

T1E12 When, under normal circumstances, may a Technician be the control operator in the Extra Class bands. **At no time**.

T1E11 Who does the FCC presume is the control operator? **The station licensee.**

T1D08 May the control operator receive compensation for operating**? Only incidental to classroom instruction at an educational institution.**

T1E05 What is a control point? **Where control operator function is performed.**

T1E09 What type of control is being used when the control operator is at the control point? **Local Control**

T1E06 Under what type of control do APRS digipeaters operate? **Automatic.**   *Hint: A digipeater is a digital repeater and repeaters are on autopilot.*

T1E08 An example of automatic control is **repeater operation.**

T1E10 What is an example of remote control? **Operating the station over the Internet.**

T1F10 Who is responsible should a repeater inadvertently transmit communications that violate FCC rules.? **The control operator of the originating station**.

IS THAT ALLOWED?

T1D06 Obscene and indecent language **is prohibited.**

T2A11 FCC rule regarding power? **Always use minimum power necessary to carry out the desired communication.**

T1A11 When is willful interference permitted? **At no time.**

T1A04 What meets the FCC definition of harmful interference? **Seriously degrades, obstructs or repeatedly interrupts.**

T1A06 **Radionavigation service** is protected from interference by amateur signals under all circumstances. *Don't cause a plane crash!*

T1D10 What is broadcasting in the FCC rules for the amateur service? **Transmission intended for reception by the general public.**

T1D12 Under what circumstances can you engage in broadcasting? **Code practice, information bulletins and emergency communications.**

T1D09 When can you broadcast assuming no other means is available? **Immediate safety of human life of protection or property.**

T1D03 When can you use codes or ciphers that hide the meaning? **Only when transmitting control commands to space stations.**

T1D04 When is the only time for music? **To manned space craft.**

T1D05 When can you notify other amateurs of radio equipment for sale or trade? **Only when the equipment is normally used in an Amateur Radio station and activity is not conducted on a regular basis.**

BANDS AND FREQUENCIES

T5C06 The abbreviation RF refers to **Radio Frequency**.

T3A07 What type of wave carries radio signals?
Electromagnetic.

T2B03 What are the two components of a radio wave?
Electric and Magnetic fields.

T3B04 How fast does a radio wave travel? **Speed of light**

T3B11 What is the approximate velocity of a radio wave?
300,000,000 meters per second.

T6D01 What is the name of current that flows in only one
direction? **Direct Current**

T5A07 What is the name for current that reverses direction on
a regular basis? **Alternating current.**

T5A12 The term that describes the number of times per
second an alternating current reverses direction is **frequency**

T3B10 **HF is 3 to 30 MHz** *High Frequency*

T3B08 **VHF is 30 to 300 MHz** *Very High Frequency*

T3B09 **UHF is 300 to 3,000 MHz** *Ultra High frequency*

T3A02 UHF works better than VHF from inside buildings
because **shorter wavelength penetrates the structure**.

T5C05 The unit of frequency is **Hertz** *(cycles per second)*.

T5B07 *Convert Megahertz (million) to kilohertz (thousand) by
moving the decimal point or changing the period to a comma..*
3.525 MHZ is 3,525 kHz

T3B01 The distance a radio wave will travel in 1 cycle is called **wavelength.**

T3B07 Different frequency bands are often identified **by their approximate wavelength.**

T3B05 How does wavelength relate to frequency? **The wavelength gets shorter the higher the frequency.**

T3B06 The formula to convert frequency to wavelength is **300 divided by the frequency in megahertz.**

T1B03 6-meter band 300/6 = frequency of 50 MHz. Answer is **52.525.** *The rest of this series of questions is answered by the closest answer.*

T1B04 146.520 MHz 300/146.52 = about **2-meter band.**

T1B07 223.50 MHZ   300/223.50 = **1.25-meter band** *closest answer.*

T1B05 70 cm band 300/.7 = **443.350 MHz** *closest answer. Notice it is 70 cm so it is .7 meters*

T1B06 23 cm band 300/.23 = **1926MHZ** *23 cm is .23 meters*

TECH PRIVILEGES

T1B10 Which bands above 30 MHz have mode restricted sub-bands? **The 6 meter, 2 meter, and 1.25 meter bands**
Cheat: There is only one answer that includes 1.25 meters.

T1B11 What emission modes are permitted in the mode-restricted sub-bands at 50 to 50.1 MHz and 144 to 144.2 MHz. **CW,** Morse Code. *Cheat: The answer is always either CW or Data.*

T1B13 Emission between 219 and 220 MHZ is **Data**   *Cheat: The answer is always either CW or Data.*

T2A10 A band plan, beyond the privileges established by the FCC is a **voluntary guideline.**

T1A02 National calling frequency on the 70cm band **446 MHz**

T1B08 Secondary use of a frequency means **you must avoid interfering.**

T1A14 If you are operating and learn you are interfering with a radiolocation service outside the United States, you must **stop operating and take steps to eliminate the harmful interference.**

T1B09 Don't get too near the edge of the band because: **Your radio might be off frequency, your transmission is wider than you think and your radio might drift as it heats up.** *All are correct.*

BE SAFE, STAY SAFE

T6A09 A **fuse** protects circuits from current overloads.

T0808 Equipment powered from a 120V AC power circuit should always include a **fuse or circuit breaker** on the hot conductor.

T0A04 A fuse **interrupts the power in case of an overload.**

T0A05 Don't put a 20 amp fuse in place of a 5 amp fuse or **excessive current could cause a fire.**

T0A06 To guard against electrical shock, use **three-wire cords, connect to a common safety ground, use ground fault interrupters.** *All are correct.*

T0A03 Green wire on a plug goes to **safety ground.**

T0B04 When putting up a tower, **look and stay clear of overhead wires.**

T0B06 Minimum safe distance from a power line? Make sure the antenna can come no closer than a **minimum of 10 feet**.

T0B09 Don't attach an antenna to a utility pole because you could **contact the high voltage lines.**

T0B02 Use a **climbing harness and safety glasses** when climbing a tower.

T0B01 Wear a hard hat and safety glasses below the tower **at all times**.

T0B03 It is **never** safe to climb a tower without a helper or observer.

T0B05 A gin pole is an extension that is used **to lift tower sections or antennas** above the top of the tower.

T0B07 Never climb a crank-up tower **unless it is fully retracted**. *Riding a tower down is very dangerous.*

T0B11 Grounding requirements are set by **local electrical codes**.

T0B12 The best conductor for RF grounding is **flat strap.**

T0B08 Proper grounding for a tower is a **separate eight-foot rod for each leg bonded to the tower and each other.**

T0B12 For lightning protection you want ground wires that are **short and direct.**

T0B10 For lightning protection, **sharp bends must be avoided**.

T0A07 Ground all the protectors to a **common plate which is in turn connected to an external ground.**

T0C04 Factors affecting RF exposure are **frequency and power level, distance from the antenna, radiation pattern of the antenna**. *All correct.*

T0C05 Exposure limits vary with frequency because **the human body absorbs more RF at some frequencies than others**.

T0C02 Which frequencies have the lowest value for Maximum Permissible Exposure Limit? **50 MHz is the worst.**

T0C03 If you run more than **50 watts at the antenna** on VHF frequencies you are supposed to do an RF exposure evaluation. *Cheat: The answer, both here and above, is "50." 50 MHz and 50 watts.*

T0C09 To stay in compliance, you should **re-evaluate whenever you change equipment.**

T0C01 VHF and UHF radio are **non-ionizing radiation**.

T0C12 RF radiation differs from ionizing radiation (radioactivity) because **RF does not have sufficient energy to cause genetic damage.**

T0C06 An acceptable way to check exposure is using **the FCC bulletin, computer modeling** or **actual measurements.** *All correct.*

T0608 To prevent exposure, try **relocating the antenna.**

T0C11 Duty cycle is the **percentage of time the transmitter is transmitting**.

T0C13 If the limit is 6 minutes and you are on 3 and off 3, you can be exposed safely for **2 times as much**.

T0C10 Duty cycle is a factor because it affects the **average exposure.**

T0C07 A person touching your antenna might get a **painful RF burn**.

RADIO OPERATION

T7A02 A transceiver **combines a transmitter and receiver** in one box.

T7A07 PTT means **Push To Talk.**

T4A01 Some microphone connectors **include PTT and voltage to power the microphone.**

T4B04 For quick access, you can store a favorite frequency in a **Memory Channel.**

T9A04 A rubber duck is a short flexible antenna. **Not as effective as full-sized**.

T9A07 You wouldn't use a rubber duck antenna in a car because **signals will be significantly weaker.**

T7A10 The device that increases the low-power output from a handheld transceiver is an **RF power amplifier.**

T4A02 Computers are used for **logging, sending and receiving CW, generating and decoding digital (data) signals** – *all correct.*

T4B02 You enter the frequency on a transceiver **by keypad or VFO knob**

T4B03 The squelch control **mutes the receiver output noise when no signal is being received.**

T2B03 Muting controlled by the presence or absence of an RF signal is called **carrier squelch**.

T2B01 If you are transmitting and receiving on the same frequency it is **simplex communication**.

T2A08 CQ is a general call for **any station**. *(Seek you?)*

T2A12 Guideline to use before calling CQ? **Listen first, ask if the frequency is in use and make sure you are in band.** *All of these are correct.*

T2A05 How do you respond to a CQ? **His call sign followed by your call sign.**

T2A09 On a repeater, in place of CQ, just throw out **your call sign** "

T2A04 To call another station on the repeater, if you know his call sign, **transmit the other station's call sign followed by your call sign**.

T3A01 If the other station reports your signals were strong a moment ago but now are weak or distorted – **move around or change the direction of your antenna.**

T3A06 Rapid fluttering when moving is called **picket fencing.**

T2B08 If two stations are on the same frequency and interfere – **common courtesy but no one has an absolute right to a frequency.**

T2B10 You are receiving interference **QRM.**

T2B11 You are changing frequency **QSY.**

T8C03 Contacting as many stations as possible during a specific time is **contesting.**

T8C04 In a radio contest you **send only the minimum information needed.**

T8C05 A Grid locater is **a letter-number assigned to a geographic location.**

EXTENDING YOUR RANGE WITH REPEATERS

T2B12 You might use simplex **if you can communicate directly without using a repeater.**

T3C10 Radio horizon is distance over which two stations can communicate on a **direct path.**

T1F09 A station that simultaneously retransmits on a different channel is a **repeater.**

T1D07 What kind of stations can automatically retransmit? **Auxiliary, repeater and space stations**. They are all repeaters.

T4B11 "Repeater offset"" is the **difference between a repeater's transmit and receive frequencies.**

T2A01 The common offset in the 2-meter band is **plus or minus 600 kHz.**

T2A03 The common offset in the 70cm-band is **plus or minus 5 MHz.**

T2B04 If you can hear but not transmit on a repeater, **you may not have the proper CTCSS, DTMF or audio tone for access**. *All are correct.*
T2B02 A sub-audible tone can be transmitted with normal voice to open the squelch of a receiver.   It is called **CTCSS** *Continuous Tone Coded Squelch.*

T1A08 Which entities recommend transmit/receive channels and other parameters for repeaters? **Frequency Coordinator**

T1A09 Who selects a Frequency Coordinator? **Amateur operators in a local area.**

T1F05 What method of call sign identification is required for a station transmitting phone signals. **CW or Phone**

EMERGENCY

T2C06 How do you get the attention of a net control station in an emergency? **Begin your transmission by saying, "Priority" or "Emergency" followed by your call sign.**

T2C07 Once you have checked into an emergency net, **remain on frequency without transmitting until asked.**

T2C09 You can operate outside of frequency privileges involving immediate **safety of human life or protection of property**.

T2C01 When do FCC rules <u>not</u> apply? **Never. FCC rules always apply.**

T2C05 RACES – **emergency and civil defense.** *All of the above.*

T2C12 ARES – emergency communications. You voluntarily **register** your qualifications and equipment.

T2C04 What do RACES and ARES have in common? **Both organizations may provide communications during emergencies.**

T2C08 Good emergency traffic handling is **passing messages exactly as received.**

T2C03 To insure a voice message is copied correctly, **use standard phonetic alphabet.**

T2C10 "Preamble" refers to the information needed **to track the message as it passes through the system.**

T2C11 "Check" is a count of **the number of words** in the message.

PROPAGATION

T5C07 Electromagnetic waves are **radio waves**.

T3A11 The **ionosphere** is the part of the atmosphere that enables propagation of radio signals around the world. *Signals bounce off the ionosphere and come back to Earth far from the point of origin.*

T2C01 Why are direct UHF signals rarely heard outside your local coverage area? UHF signals **do not reflect off the ionosphere.** *They go right through it which is why UHF is often used to talk to satellites.*

T3C11 Why do VHF and UHF signals usually travel further than visual line of sight? **The Earth seems less curved to radio waves than light.** *The answer seems so bizarre, it must be right. Actually air bends the radio wave a bit over the horizon.*

T3C05 **Knife-edge diffraction** might cause radio signals to be heard despite obstructions. *Waves bend around a sharp corner.*

T2C06 VHF and UHF of 300 miles is possible because of **tropospheric scatter.**

T2C08 What causes tropospheric ducting (scatter)? **Temperature inversions** trap and bend the signal.

T3C03 Auroral reflection **exhibit rapid fluctuations of strength and often sound distorted.**

T3C07 Meteor scatter is usually on **6 meters**.

T3C02 VHF signals are received from long distances due to **refraction from a sporadic E laye**r.

T3C04 Strong over-the-horizon signals on 10,6 and 2-meter bands are also **sporadic E.**

T3C09 The best time for long-distance 10 meter propagation is **after dawn until sunset during high sunspot activity**. *Daylight and high sunspots.*

T3C12 Long distance communication at the peak of the sunspot cycle are on **6 or 10 meters.**

SATELLITES

T8B01   You can talk to a satellite as long as your license allows you to operate on the **satellite's uplink frequency**.

T8B04 You can talk to the International Space Station if you have a **Technician or above** license.

T8B10 LEO tells you the satellite in in **Low Earth Orbit.**

T8B03 Satellite tracking programs provide maps showing **the position of the satellite, time, azimith and elevation and the Doppler shif**t. *All answers correct.*

T8B06 The inputs to a tracking program are the **Keplerian elements.**

T8B05 A satellite beacon is a transmission from space that **contains information about a satellite.**

T8B09 Spin fading is because of **rotation of the satellite and antennas.**

T8B07 Doppler shift is an observed change in frequency caused by the **relative motion between the satellite and Earth station.**

T8B08 A satellite operating in U/V mode is using an uplink in UHF and a downlink in VHF. Answer: **70cm and 2-meter band**

T8B02 Transmitter power on the uplink frequency should be the **minimum amount of power to complete the contact.**

T8B11 For sending signals, digital satellites commonly use **FM Packet.**

T1A07 Telemetry is one-way **transmissions of measurements.**

T1A13 Telecommand is a one-way transmission to **initiate, modify or terminate functions.**

COMPUTERS AND HAM RADIO

T8D01 Digital communications can be **PSK31, Packet, MFSK.** *All are correct.   The computer decodes the digital signal and shows it as typed letters on your screen.*

T4A06 Connected between a transceiver and computer in a packet radio station is a **TNC – terminal node controller.** *A modem.*

T4A07 A computer sound card can send **sound to the microphone and converts received audio** to digital form.

T8D08 Packet uses **headers, checksums and error correcting.** *All are correct.*

T8D11 ARQ is system to **detect errors and request retransmission.** *Automatic Repeat Query.*

T3A10 Multiple paths are **likely to increase the error rates.** *If signals take different paths and arrive at slightly different times, they confuse the decoding.*

T8D05 APRS Automatic Packet Reporting System works **in conjunction with a map showing the location of stations**.

T8D02 APRS means **Automatic Packet Reporting System**. *Your radio broadcasts your position and you show up on a computer map.*

T8D03 Data to the transmitter for automatic position reports is supplied by a **GPS Global Positioning System receiver** to provide data to the transmitter.

T8D05 PSK is **Phase Shift Keying.**

T8D07 PSK31 is a **low rate data transmission mode**.

T8C11 An **Internet Gateway** connects a radio to the Internet.

T8C13 Internet Radio Linking Project (IRLP) **connects a repeater to the Internet.**

T8C06 Access an IRLP node by using **DTMF signals.** *The touchtone buttons on your microphone dial the number where you want to go.*

T8C10 You select a specific IRLP node using **the keypad to transmit the IRLP node ID.** *Hook into an IRLP node and dial 5600, your signal will travel over the Internet and pop out on a repeater in London.*

T8C12 Voice Over Internet protocol is a **method of delivering voice over the Internet.** *Skype is a use of VoIP.*

T8C09 You can find lists of active nodes that use VoIP from a **repeater directory.** *There's an APP for that or you can buy a book.*

T8D04 **NTSC** is a **color TV signal.** *Cheat: Think C for color*

T8A10 Fast scan TV is about **6 MHz wide.** *Cheat: Remember TV6*

MODES

T7A09 If you are going to do weak signal work, you need a **multi-mode transceiver** *so you can operate modes other than FM. FM requires a stronger signal.*

T7A08 Combining speech with an RF carrier is called **Modulation.**

FM FREQUENCY MODULATION

T8A04 Repeaters use **FM modulation.** *FM modulation varies the frequency of the radio wave to carry the voice or data.*

T8A09 The approximate bandwidth of an FM phone signal is between **10 and 15 khz** wide. *That is how much the frequency varies or "deviates."*

T2B05 Amount of deviation is determined by **amplitude** (loudness).

T2B06 Increased deviation means **the signal occupies more bandwidth.**

T8A02 VHF Packet (data bursts) run on **FM.**

SSB SINGLE SIDEBAND MODULATION

T8A01 Single Side Band is a form of **amplitude modulation.**

T8A03 Long distance weak signal VHF/UHF contacts are most often **single sideband (SSB).**

T8A07 A primary advantage of single sideband over FM is **SSB signals have narrower bandwidth.** *Three SSB signals can fit in the space on one FM.*

T8A08 The bandwidth of a single sideband voice signal is **3 kHz.** *Cheat: SSB has 3 letters*

T4B09 The filter bandwidth to minimize noise and interference to SSB **is 2400 Hz.** *The closest choice to the signal bandwidth.*

T8A06 Which sideband on 10 meters HF, VHF and UHF? **Upper** *Cheat: Remember upper frequency = upper sideband. Upper is always the test answer.*

T2B13 SSB above 50 MHz is **permitted in at least some portion of all the amateur bands above 50 MHz.**

T4B06 If the voice pitch seems too high or low use the **receiver RIT or clarifier.**

T4B07 (RIT) means **Receiver Incremental Tuning.** *Fine tuning.*

CW MORSE CODE

T8A05 The type of emission that has the narrowest bandwidth is **CW.** *CW means "continuous wave. It is not modulated – it is either on or off.*

T8D09 Code used with CW is **International Morse Code.**

T8D10 CW can be sent with **a key, keyer or keyboard** *All are correct.*

T8A11 The bandwidth for CW is **150 Hz.**

T4B10 The appropriate filter bandwidth to minimize noise and interference for CW is **500 Hz.** *The closest choice to the signal bandwidth.*

T4B08 The advantage of having multiple bandwidth choices is it permits noise or interference reduction by **selecting a bandwidth matching the mode.**

RADIO DESIGN

T7A05 The circuit that generates a signal at a desired frequency is called an **oscillator**.

T7A01 The presence of a signal is determined by the receiver's **sensitivity**.

T7A04 The ability to discriminate between signals is called **selectivity**.

T7A03 You convert a signal from one frequency to another with a **mixer**.

T7A06 You convert a low powered 28 MHz exciter to 222 MHz output using a **transverter**. *Cheat: If you see "transverter" it is always correct.*

T4B12 AGC – automatic gain control **keeps the receive audio relatively constant**. *So you don't blow out your ears tuning across a loud station.*

T7A11 You would put an RF preamplifier **between the antenna and receiver**. *It is "pre" or before the radio.*

T4A03 A regulated power supply **prevents voltage fluctuations from reaching sensitive circuits**.

CLEAN UP THE SIGNAL

T7B10 If your audio is distorted or unintelligible, the problem could be caused by **location, low batteries** or being **off frequency**. *All correct.*

T4B01 Too high a microphone gain and **the output signal may become distorted**.

T7B01 If you are told you are over-deviating, **talk further away** from the microphone.

T2B07 Your FM signal might interfere with other stations if your **microphone gain is too high causing overdeviation**. *Your signal is too wide.*

T4B05 Ignition noise might go away if you turn on the **noise blanker**.

T4A10 A high pitched whine that varies with speed is usually the **alternator** of your car getting in the radio.

T4A12 Variable high-pitched whine on the audio is **noise on the vehicle's electrical system being transmitted**. *Really the same question as above.*

T4A09 To cure distorted audio caused by RF current on the shield of a cable use a **ferrite choke.**

T7B12 The first step to resolve cable TV interference is to **be sure all the connectors are properly installed**.

T4A04 To reduce harmonic emissions install a filter **between the transmitter and antenna** to control harmonic emissions. *You want to stop the harmonics before they get to the antenna.*

T7B03 Radio frequency interference can be caused by **fundamental overload, harmonics, and spurious emissions**. *All are correct.*

T7B11 A symptom of RF feedback is reports of **garbled, distorted or unintelligible transmissions.**

T6D12   Shielded wire **prevents coupling of unwanted signals to or from the wire.**

T7B06 If a neighbor complains you are interfering with their radio or TV, **make sure your station is not causing interference to your own radio or television.** *First, check your own gear.*

T7B02 An AM/FM radio might receive Amateur Radio transmissions because the receiver is **unable to reject strong signals outside the AM or FM band.** *An example of fundamental overload.*

T7B04 To reduce or eliminate interference to a nearby telephone, **put an RF filter on the telephone.** *The phone line is acting as an antenna and you need to choke off the RF.*

T7B07 To correct a radio frequency interference use **low and high pass filter, band reject and band pass filters, ferrite chokes.** *All are correct*

T7B05 To reduce the overload of a non-Amateur Radio or TV, block the amateur signal with a filter on the **antenna input of the receiver.**

T7B09 Part 15 devices are **unlicensed and may emit low power radio signals.** *(modems, cordless phones weather stations, wireless printers).*

T7B08 If something in a neighbor's home is causing harmful interference to your amateur station, **work with your neighbor, check your station, politely inform.** *All the choices are correct.*

ELECTRONIC THEORY

VOLTS

T5A05 Electromotive force is called **Voltage.**

T5A11 The basic unit the **volt.**

T7D01 You would measure electromotive force with a **voltmeter.**

T7D12 If you are measuring high voltages **ensure that the meter and leads are rated for use at the voltages to be measured.** *Or the high voltage can zap you through the insulation.*

T7D02 You connect a voltmeter in **parallel with the circuit.** *Across the circuit not in line with it.*

AMPERES

T-5A03 Flow of electrons is **current.**

T5A01 Electric current is measured in **amperes.**

T7D04 Current is measured with an **ammeter.**

T0A02 Current flowing through the body can cause a health hazard by **heating tissue, disrupts the electrical functions of cells, causes involuntary muscle contractions.** *All are correct.*

T7D03 An ammeter is connected **in series with the circuit.** *You are measuring current flowing and to do that you must be in line with the circuit.*

RESISTANCE

T5A07 A good electrical conductor is **copper.**

T5A08 A good electrical insulator is **glass**.

T6A01 The component that opposes the flow of DC current is **resistor**.

T7D05 You measure resistance with an **Ohmmeter**. *Measured in ohms.*

T6A02 An adjustable volume control is often a **potentiometer**. *Variable resistor.*

T6A03 A potentiometer controls **resistance.**

INDUCTORS

T6A06 An **inductor** stores energy in a magnetic field.

T6A07 Inductor is composed **of a coil of wire** *That is why it is sometimes called a "coil."*

T5C03 The ability to store energy in a magnetic field is called **inductance**.

T5C04 The basic unit of inductance is the **henry**.

T5C12 Impedance is the **measure of opposition of AC current flow in a circuit.** *Think of that magnetic field building up and reversing. The collapsing magnetic field opposes the change of direction.*

T5C13 Units of impedance are **ohms**. *Just like resistance only this is resistance to AC.*

CAPACITORS

T5C01 The ability to store energy in an electric field is called **capacitance.**

T7D10 If an Ohmeter *(which measure resistance)* indicates low resistance and then increasing resistance with time, **the circuit contains a large capacitor.** *It is charging up.*

T6A04 The electrical component that stores energy in an electrical field is called a **capacitor.**

T5C02 The basic unit of capacitance is called a **farad.** *Named after Michael Faraday.*

T6A05 Two or more conductive surfaces separated by an insulator, make up a **capacitor.**

T0A11 When a power supply is turned off you might still **receive a shock from stored charge in a large capacitor.**

TRANSISTORS

T6B03 A **transistor** can be used as a switch or an amplifier

T6B01 Components capable of using voltage or current to control current flow? **Transistors.**

T6B05 Electronic components that can amplify signals? **Transistors**.

T6B12 Term that describes a transistor's ability to amplify a signal is called **gain.**

T6B10 The three electrodes on a transistor are **emitter, base and collector** *Cheat: All three words are unique in the answers. Pick one to remember or remember EBC.*

T6B04 A component made of three layers of semiconductor material is a **transistor.** *Transistors are sometimes called semi-conductors.*

T6B08 An FET is a **Field Effect Transistor.**

T6B11 An FET has a **source, gate and drain.** *Cheat: A field has a gate and a drain, get it?)*

DIODES

T6B02 The component that allows current to flow in only one directions is a **diode.** *Diodes convert Alternating Current to Direct Current.*

T6B09 The two electrodes on a diode are **anode and cathode.** *Annie and Cathie.*

T6B06 The cathode end of a diode is **identified with a stripe.**

VOLTS, OHMS, AMPERES & POWER – OHM'S LAW

Voltage = Amps times Resistance

Voltage is symbolized by the letter E,
Amps the letter I,
Resistance the letter R.

*Memorize: VoltagE, **R**esistance and the other one Amps, **I**.*

So the formula looks like E=IR.

How do you remember it? Use the magic circle.E=IR: Eagle flies over the Indian and the Rock.

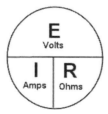

Put your thumb over the value you are solving for and the formula is what is left. For example, if you are solving for R, cover up the R and the answer is E/I, voltage divided by

amperage.   Cover the answer and see if you can do it. Don't panic. Take your time.

T5D04 What is the resistance of a circuit in which a current of 3 amperes flows through a resistor connected to 90 volts?
90/3 = 30 ohms

T5D05 What is the resistance in a circuit for which the applied voltage is 12 volts and the current flow is 1.5 amperes?
12/1.5 = 8 ohms

T5D06 What is the resistance of a circuit that draws 4 amperes from a 12-volt source?

12/4 = 3 ohms

T5D07 What is the current flow in a circuit with an applied voltage of 120 volts and a resistance of 80 ohms?
120/80 = 1.5 amperes

T5D08 What is the current flowing through a 100-ohm resistor connected across 200 volts?
200/100 = 2 amperes

T5D09 What is the current flowing through a 24-ohm resistor connected across 240 volts?
240/24 = 10 amperes

T5D10 What is the voltage across a 2-ohm resistor if a current of 0.5 amperes flows through it?
2 x .5 = 1 volt

T5D11 What is the voltage across a 10-ohm resistor if a current of 1 ampere flows through it?
10 x 1 = 10 volts

T5D12 What is the voltage across a 10-ohm resistor if a current of 2 amperes flows through it?
10 x 2 = 20 volts

POWER IN WATTS

T5A10 The rate at which energy is used is called **power**.

T5A02 Electrical power is measured in **Watts**.

T5C08 What is the formula to calculate electrical power in a DC circuit? Power equals **voltage multiplied by current**. *It is easy as PIE, P=IE and of course, we have another magic circle to help us.*

Cover the answer and see if you can do it. Don't panic. Take your time.

T5C09 How much power is being used in a circuit when the applied voltage is 13.8 volts DC and the current is 10 amperes? **13.8 x 10 = 138 watts**

T5C10How much power is being used in a circuit when the applied voltage is 12 volts DC and the current is 2.5 amperes? **12 x 2.5 = 30 watts**

T5C11 How many amperes are flowing in a circuit when the applied voltage is 12 volts DC and the load is 120 watts? **120/12 = 10 amperes**

HOW TO DRAW A RADIO

T6C01 **Schematic symbols** are standardized representations of electrical components.

T6C12 The symbols in an electric circuit diagram represent **electrical components.**

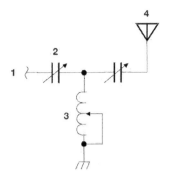

Figure T-3

T6C10 *Look at Figure T-3.* Component 3 looks like a coil and it is a coil (inductor). *Because it has an arrow pointed to it, it is a* **variable inductor.** *Imagine the arrow sliding up and down to change the value.*

T6C11. Component 4 looks like a funnel to collect radio waves. It is an **antenna.**

T6D08 A tuned circuit is composed of an inductor and a **capacitor.** *Cheat: tuned circuit = inductor and capacitor.*

T6D11 A simple resonant or tuned circuit is an **inductor and capacitor that form a filter.**

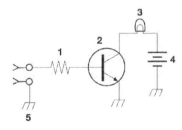

Figure T-1

T6C02 Looking at Figure T-1, component 1, the zigzag line is a **resistor.** *Imagine the resistance if you had to run a zigzag line like that.*

T6C03 Component 2 is a **transistor**. *Three leads: emitter, base and collector.*

T6D10 The function of the transistor is **to control the flow of current.** *It can be an amplifier or a switch.*

T6A08 A **switch** connects or disconnects circuits.

T6C04 Figure 3, a **lamp** *looks like a lamp.*

T6C05 Component 4 is a **battery** *with long and short lines representing the cells.*

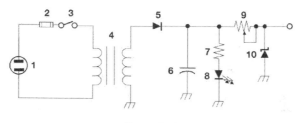

Figure T-2

T6D03 Figure T-2, Component 3 is a switch. The one in the diagram is **single-pole single-throw**. *One wire in and one wire out*

T6C09 Component 4 is a **transformer.** *See the different sets of windings that change voltages from one side to the other? The ratio of the number of windings determines the step-up or step down result.*

T6D06 What component is used to change 120v AC house current to lower voltage for other uses? **Transformer**.

T6C06 Component 6 is a **capacitor**. *A capacitor has two layers. If there are more than two layers, you are looking at the symbol for a battery.*

T6C07 Component 8 is a **light emitting diode**. *Think of the arrows as light coming out of the LED.*

T5B07 LED means **light emitting diode.**

T6d07 A **light emitting diode** is commonly used as a visual indicator. *The buttons that light up on your TV remote are LEDs.*

T6C08 Component 9 is a **variable resistor.** *Zigzag line makes it a resistor, the arrow makes it variable.*

T6D04 A **meter** is used to display signal strength

T6D02 A relay is a **switch controlled by an electromagnet.**

T6D05 A **regulator** controls the voltage from a power supply. *It regulates.*

T6D09 An **integrated circuit** combines several semiconductors in one package.

DECIBELS

T5B11 Going from 20 to 200 watts is 10 times or 10 dBs. *But since we are talking about logarithms here, it is not linear.*

T5B09 Going from 5 watts to 10 watts is **3 dB**. *Doubling is 3 dB. It doesn't matter where you start, if you double, it is 3dB, and if you double again it is 6dB. If you halve the power it is -3dB, halve it again and it is -6dB.*

T5B10 A power decrease from 12 watts to 3 watts is minus half and half again **-6dB.**

MOVING DECIMALS

T5B02 1,500,000 hertz is the same as **1500 KHz**. *We are usually dealing with big numbers so it is convenient to leave out all the zeroes. Kilo means thousand so 1500 KHz is 1500 thousand hertz or 1,500,000 hertz or 1.5 MHz – they all say the same thing.*

T5B13 2425 MHz would be the same thing as **2.425 Ghz**. *Gigahertz. Million million hertz. When going to a larger identifier, put in a comma, then change it to a period.*

T5B12 28,400 Khz would be the same as **28.400 MHz** *Put in a comma, if there is not one already, and then change it to a period.*

T5B03 A kilovolt would be **one thousand volts.** *Kilo is thousand.*

For really small numbers we use, milli (1/1000) and micro (1/,1,000,000.)

T5B04 A microvolt is a millionth of a volt. *Micro is millionth.* So one microvolt volt is **one one-millionth of a volt.**

T5B06 A 3000 milliampere current would be 3000 /1000 = **3 amperes.**

T5B05 500 milliwatts would be 500/1000 = **0.5 watts.**

T5B01 1.5 amperes would be how many thousandths? 1.5 X 1000 = **1,500 milliamperes.**

T5B08 Picofards are really tiny: one millionth of one millionth of a farad. So 1,000,000 of them = 1 millionth of a farad or **1 microfarad** *I don't blame you if you have to memorize this one.*

BUILDING EQUIPMENT AND MEASURING VALUES

T7D08 Radio and electronic soldering uses **rosin core** solder. *Acid would eat away the connection.*

T7D09 A cold solder joint is **grainy and dull.**

T7D07 Multimeters measure **voltage and resistance.**

T7D11 If you're measuring resistance, be sure the circuit is **not powered.**

T7D06 You'll damage the meter if you try to **measure voltage when using the resistance setting.**

ANTENNAS

POLARIZATION

T9A10 Radiation is strongest **broadside to the antenna.**

T9A03 A simple dipole mounted parallel to the Earth's surface is **horizontally polarized**. *The wire is horizontal and the radiation is strongest broadside so the radiation is also horizontal.*

T3B02 What property is used to describe a radio wave's polarization. The **electrical field orientation**.

T3A03 The antenna polarization used for CW and SSB is usually **horizontal**. *Mobile FM is vertical because the antenna fits on the car that way.*

T9A02 A vertical antenna has an **electrical field perpendicular** to the Earth.

T9A13 Mounting the antenna in the center of a vehicle roof provides the most **uniform radiation pattern**.

T3A04 If you are using different polarizations, signals are **significantly weaker**.

T3A09 Skip signals are elliptically polarized and you can use either a **vertically or horizontally polarized antenna**.

T3A08 Irregular fading can be caused by **random combining of signals arriving from different paths**.

ANTENNA LENGTHS

T9A09 A half-wavelength antenna is half a wavelength long. 6 meters / 2 = 3 meters long for a half wave. That is about 9 feet x 12 = **112 inches** is the closest answer.

T9A08 A quarter-wave for 146 MHz. **19 inches**. *Divide 234 by the frequency = the length of a quarter-wave in feet.*

T9A05 To make an antenna resonant on a higher frequency, you would **shorten it**. *Higher frequency, shorter wavelength.*

T9A12 A 5/8 wave antenna offers **a lower angle of radiation** and more gain than a ¼ wave.

T9A14 A type of loading can be an **inductor to make an antenna electrically longer**.

DIRECTIONAL ANTENNAS

T9A06 Yagis, Quads and Dishes are all **directional antennas**.

T9A01 A beam antenna **concentrates signals in one direction**.

T9A11 The gain of an antenna is **the increase in signal strength in a specific direction compared to a reference antenna.**

T8C01 You could locate the source of interference or jamming with **radio direction finding.**

T8C02 A directional antenna is also useful for a **hidden transmitter hunt**.

T3A05 When using a directional antenna, how might your station be able to access a distant repeater if buildings or obstructions are blocking the direct line of sight path? Turn your antenna to **try to find a path that reflects signals to the repeater**

COAXIAL CABLES

T7C12 A common use of coaxial cable is carrying **RF signals between a radio and antenna**.

T9B03 Why is coax used more often that other feed line? Coax is **easy to use and requires few special installation considerations.**

T9B02 Impedance of most commonly used coaxial cable is **50 Ohms.**

T9B05 When increasing frequency, coaxial cable **loss increases.**

T9B07 A PL 259 connector is commonly used on **HF** frequencies.

T9B06 Frequencies above 400MHz use a **Type N** connector

T9B11 The lowest loss at VHF would be **air-insulated hard line.**

T7C11 Air core coax has the lowest loss at high frequencies but the disadvantage is **requires special techniques to avoid water absorption**.

T7C09 The most common cause for failure of coaxial cable is **moisture contamination**.

T9B08 Seal the connectors to prevent an **increase in feed line loss** *(from moisture contamination)*.

T7C10 Coax outer jacket should be resistant to ultraviolet light because ultraviolet light from the sun can **damage the jacket** and allow water to enter.

T9B10 Larger RG-8 cable will have **less loss** than smaller RG58.

ANTENNA ANALYZERS AND SWR

T7C02 We use an **antenna analyzer** to determine if the antenna is resonant.

T7C03 The standing wave ratio (SWR) is a measure of **how well the load is matched**. *Ratio of forward vs reflected power.*

T4A05 You connect an SWR meter in series **between the transmitter and the antenna**. *Connect it in line.*

T9B01 You want a low SWR **to allow the efficient transfer of power and reduce losses.**

T7C04 A perfect match is an SWR of **1 to 1**.

T7C05 Most radios will automatically reduce power if they sense an **SWR above 2 to 1.**

T7C06 An SWR of 4:1 would indicate an **impedance mismatch.**

T9B09 If have erratic changes in SWR readings, you probably have a **loose connection in the antenna or feed line.**

T7C08 You can also measure proper match with a **directional wattmeter**.

T7C07 Power lost in the feed line is **converted to heat.**

T9B04 You can use an antenna tuner to help **match the antenna system impedance to the transceive**r.

T7C01 A dummy load **prevents radiation of signals when making tests.**

T7C13 A dummy load consists of a **non-inductive resistor and heat sink.**

BATTERIES

T6A10 Rechargeable batteries include **Nickel-metal hydride, Lithium-ion and Lead-acid gel-cells.** *All are correct.*

T6A11 Non-rechargeable batteries are **Carbon Zinc.**

T5A06 A mobile transceiver requires **about 12 volts**

T2C02 To recharge a 12 volt lead-acid battery if the power is out, **connect it in parallel with a vehicle's battery and run the engine.**

T0A10 If you try to charge or discharge a lead-acid battery too quickly, **it can overheat and give off flammable gas or explode.**

T0A01 A safety hazard of 12-volt storage batteries is that **shorting the terminals can cause burns, fire or an explosion.**

T0A09 A hazard presented by a conventional 12-volt storage battery is that **explosive gas can collect if not properly vented.**

T4A11 Connect the negative side of the power cable to the **battery or engine block ground strap.**

~~~~~~

INDEX

AGC, 67, 117

Anode, 75, 123

Amperes, 13, 82, 90, 131, 135, 136, 137, 141

APRS, 66, 100, 114

ARES, 60, 61, 110

ARQ, 65, 114

ARRL, 7, 52

Auroral reflection, 44

Automatic repeat request, 65, 114

Base, 36, 74, 80, 122, 127

Battery, 12, 32, 69, 81, 82, 94, 95, 127, 128, 134, 135

Beacon, 62, 112

Beam antenna, 88, 89, 131

Broadcasting, 27

Capacitor, 69, 73, 74, 80, 81, 82, 122, 126, 128

Cathode, 75, 123

Check, 7, 61, 65, 72, 111

Checksum, 65

Club, 23, 55, 61, 65, 99

Coaxial cable, 89, 90, 132, 133

Collector, 74, 80, 122, 127

Control operator, 26, 29, 30, 31, 99, 100

Control point, 29, 30, 100

CQ, 53, 58, 59, 108

CSCE, 7

CTCSS, 56, 109

CW, 1, 3, 18, 24, 36, 39, 40, 41, 43, 69, 86, 103, 107, 110, 116, 117, 130

Deviation, 37, 69, 71, 115

Digipeater, 66, 100

Digital communications, 36, 65

Diode, 75, 82, 123, 128

Directional antenna, 14, 83, 89, 132

Doppler shift, 63, 112, 113

DTMF, 56, 57, 109, 114

Dummy load, 93, 134

Duty cycle, 49, 106, 107

Emitter, 74, 80, 122, 127

Farad, 74, 84, 122, 129

Ferrite choke, 70, 118

FET, 75, 122, 123

Filter, 39, 40, 69, 70, 71, 72, 74, 80, 116, 117, 118, 119, 126

Frequency, 21, 32, 33, 37, 48, 56, 57, 97, 102, 110

Frequency Coordinator, 56, 110

Fundamental overload, 71, 118, 119

Grid locater, 52, 109

Ground, 46, 47, 48, 88, 90, 95, 104, 105, 106, 135

Harmonics, 71, 118

Header, 65

Henry, 73, 121

Ignition noise, 67, 118

Impedance, 73, 121, 132

Inductance, 73, 81, 121

Integrated circuit, 82, 128

Internet Gateway, 57, 114

IRLP, 57, 114

ITU, 18, 20, 22, 28, 35, 97

Keplerian elements, 63, 112

Knife-edge propagation, 43

Lead-acid, 12, 94, 134

LED, 82, 128

LEO, 62, 112

Meteor scatter, 14, 44

MFSK, 36, 40, 65, 113

Military station, 28

Mixer, 67, 117

Modulation, 36, 115

Morse Code, 16, 18, 24, 39, 62, 95, 103, 116

Multi-path, 42, 66, 69, 87

Music, 27

NTSC, 36, 115

Oscillator, 50, 67, 117

Packet, 36, 37, 64, 65, 66, 113, 114, 115

Part 15, 72, 119

PART 97, 20

PL 259, 91, 132

Polarization, 86, 130

Preamble, 61, 111

Preamplifier, 68, 117

PSK31, 36, 40, 65, 66, 113, 114

PTT, 50, 107

QRM, 51, 108

QSY, 52, 108

RACES, 60, 61, 110

Regulator, 82, 128

Relay, 54, 82, 128

Repeater, 31, 32, 37, 43, 45, 53, 54, 55, 56, 57, 58, 60, 61, 62, 64, 66, 87, 89, 100, 108, 109, 114, 115, 132

Resistance, 13, 73, 76, 77, 123

RF feedback, 70, 119

RIT, 39, 116

Rubber duck, 51, 107

Scatter mode, 44 ,111

Schematic symbols, 79, 125

Selectivity, 39, 67, 117

Semiconductor, 74, 122

Simplex, 50, 54, 55, 58, 108, 109

Spin fading, 63, 113

Spurious emissions, 71

Squelch, 51, 56, 107, 108, 109

SSB, 1, 2, 38, 39, 40, 65, 86, 115, 116, 130

SWR, 2, 3, 91, 92, 93, 133, 134

Telecommand, 62, 113

Telemetry, 62, 113

TNC, 65, 66, 113

Transformer, 81, 82, 127

Transistor, 74, 75, 80, 81, 122, 127

Transverter, 67, 117

Tropospheric ducting, 44

Type N connector, 91, 132

U/V mode, 64, 113

Voltage, 12, 76, 77, 78, 120, 123

Watts, 78, 125

Wavelength, 33, 87, 102, 103, 131

Made in the USA
Middletown, DE
06 May 2017